SURVIVAL OR EXTINCTION

Proceedings of a Conference held at the
Royal Botanic Gardens, Kew, entitled

The Practical Rôle of Botanic Gardens
in the Conservation of Rare and Threatened Plants

11–17 September 1978

Edited by

Hugh Synge and Harry Townsend

1979
Published by The Bentham-Moxon Trust
Royal Botanic Gardens, Kew

ISBN 0 9504876 2 7

Typeset in 10 on 11 pt Baskerville and printed by
Unwin Brothers Ltd.,
The Gresham Press, Old Woking, Surrey

Cover design by W. Jennison, Royal Botanic Gardens, Kew

Acknowledgements

It is a great pleasure to take this opportunity, on behalf of the Royal Botanic Gardens, Kew, to thank all those who made this Conference possible. Initially our thanks are due to the members of the Steering Committee set up by the previous Kew Conservation Conference in 1975. The group consisted of Mr C. D. Brickell, Professor K. Esser, Professor J. G. Hawkes, Mr D. M. Henderson, Professor V. H. Heywood and Dr S. M. Walters, as well as Mr G. Ll. Lucas, Mr R. I. Beyer and myself from the Kew staff. This committee funded the follow-up work to the 1975 conference, in particular the tri-lingual background paper by Dr Walters in *Gärtnerisch-Botanischer Brief* 51 (1977) and the TPC European Botanic Gardens Conservation Project (see pp. 21–22). In July 1977 this committee set up a Working Party to plan the present conference, with the aim of providing practical guidance, primarily for technical managers, on the rôle of botanic gardens in plant conservation. This Working Party, with myself as Convenor, consisted of Professor Esser, Mr Brickell and Dr Walters with Mr Lucas, Mr Beyer and Mr D. W. H. Townsend from the Kew staff.

The major part of the detailed organization of the conference was undertaken by Mr Townsend, who was appointed Conference Director. His intense activity ensured the smooth running of the whole conference and I would like to express my particular thanks to him.

That the conference was so successful and was recognized as of great value by delegates was due in large measure to the hard work of all those who were involved. Particular thanks are due to A. M. Chabert, Conference Treasurer, and to I. E. Byrne and G. E. Maryon, who provided the Conference Secretariat. The reception team at Kew and at the Froebel Institute, where delegates were housed, consisted of M. M. Arnold. N. R. Brown, A. M. Chabert, D. Emanuel, R. P. Harrison, V. Horwill, W. Jennison, G. E. Maryon, J. G. Plummer and G. E. Smith. Meeting the delegates at London Airport was undertaken by G. H. Cobham, B. M. Dickinson, J. Dixon, S. Goodenough, A. Mayo, D. R. Owen, G. A. Pattison, P. R. Potter and P. M. Rutherford. Discussions were recorded by G. L. R. Bromley, A. M. Chabert, G. H. Cobham, H. Fliegner, P. L. Gibbon, W. Jennison, J. Lonsdale, D. Mason, K. L. Miller, D. R. Owen and A. H. M. Synge, under the leadership of N. P. Taylor. Exhibits were arranged by R. C. R. Angel, P. Reid, L. Giuffrida, J. M. Ruddy and R. Strong. Photography and projection were undertaken by T. A. Harwood, L. A. Pemberton and R. R. Zabeau. The conference dinner was arranged by A. M. Chabert, P. R. Potter and C. Sawyer. Administration was by S. Brookes, M. A. Arnold-Gilliat, F. A. Ball and H. M. Hyde. To all these I would like to express our grateful thanks on behalf of the organizing committee.

The Royal Botanic Gardens, Kew, would also like to express their appreciation to the following organizations for their help in ensuring the success of the conference:

The NATO Scientific Affairs Division
British Caledonian Airways
The British Council
The Commonwealth Foundation
The Forestry Commission and Mr A. Westall
Hampshire County Council
The Incorporated Froebel Institute
The National Trust
The Nature Conservancy Council
Queen Elizabeth Countryside Park
The University Botanic Garden, Cambridge

The progress of the various sessions owed much to the Conference Chairmen who willingly took on this extra rôle, and we would like to thank Dr T. Lasser, Dr G. W. M. Barendse, Mr A. Brady, Mr A. Thompson, Professor G. Moggi, Mr A. R. Hassan King, Dr C. D. Adams, Mr P. Chai and Dr G. T. Prance, for their help and guidance.

After the Conference, responsibility for the Proceedings was borne by a small Editorial Panel of Mr Beyer, Mr Lucas and myself, supporting the two editors, Mr Synge and Mr Townsend, who bore the main brunt of editing and producing the book. Mr D. V. Field kindly undertook the preparation of the index and J. Vasek kindly translated L. V. Asieshvili's paper into English. I would here also like to thank M. J. Brind, E. N. Attwood, J. A. Curtis, E. J. Fitchett, J. L. Ronald, E. Kingston and A. M. Beyer for the endless typing necessary in the preparation of such a work, the staff of the Library, in particular D. C. Scott, J. M. Whyte and P. J. Atkin for their help over editorial matters, and C. M. Grey-Wilson and C. T. Town from the Conservation Unit for their essential support work.

Even with this long list not everyone is mentioned by name but nevertheless the work of all contributors was recognized and appreciated by the delegates for whom we hope the way to a more active conservation policy has now been made more clear.

J. B. Simmons
Curator

Foreword

This book comprises the lectures and papers presented to the second conservation conference at Kew. The conference aimed to highlight the crucial rôle of botanic gardens in plant conservation and in this it was conspicuously successful. Delegates must surely have returned home stimulated by the ideas—and increasingly the realities—of what is being undertaken around the world. Establishing a new garden with only minimal resources (A. P. Vovides), undertaking a joint programme with botanic gardens and conservation bodies to conserve an entire country's flora (Sven Wahlberg), developing ideas for 'evacuating' critically endangered species to rehabilitated ecosystems (A. V. Hall & H. B. Rycroft)—in numerous ways the conference pointed the road ahead as well as proving that success is possible.

One principle emerges clearly: conservation in the wild is nearly always the most desirable policy. Indeed, the conservation activities of botanic gardens are most successful when directed to that end, as for example in helping to identify threatened plants and recommend appropriate reserves (E. E. Gogina), in alerting the public to the need for plant conservation and the pleasure that plants can give (Don Aldridge), or even in holding small reserves themselves (David Bramwell). As Sir Peter Scott—Chairman of the Survival Service Commission of the International Union for Conservation of Nature and Natural Resources (IUCN)—has repeatedly emphasized for both threatened plants and threatened animals, it is their habitats and ecosystems that must be conserved.

Special mention should be made of the need for more botanic gardens in the tropics and for more support—facilities and finance—for those already in existence. The conference papers add up, *in toto*, to a powerful plea for plant conservation. They will equip those whose rôle it is to try to extract funds for this purpose from governments and funding agencies with excellent supporting arguments. The money required is not large in international terms and some encouragement is badly needed, if only to boost the relatively small but highly valuable sums that charities like the World Wildlife Fund are able to provide.

This conference should in no sense be seen as a charter to collect rare and vanishing species indiscriminately. Collaboration is urgently needed to ensure botanic gardens know which species are threatened in the wild so that they can especially treasure and build up those they already have for distribution to other gardens. It was for this reason that the Conference requested the Threatened Plants Committee of the Survival Service Commission of IUCN to set up a Botanic Gardens Conservation Co-ordinating Body with a brief to find out which threatened plants are in cultivation and where, and to keep botanic gardens informed of current conservation activities. Already a secretary and treasurer have been appointed, a draft programme has been agreed, and initial projects are underway.

But information-gathering and discussion are not enough. Rather they are the starting point for action. No-one can doubt the severity of the threats to the plant kingdom, nor remain complacent about the massive loss of species that are likely in some tropical regions. However, there are grounds for hope. There is increasing awareness of the need for plant conservation as an integral part of man's long-term welfare and survival. The activities and projects described in this book are already achieving success—though virtually all of them could be rapidly developed and expanded if adequate support were forthcoming.

All animal life, man most certainly included, is entirely dependent on the plant kingdom for continued existence. No plant life, no animal life.

As gene banks, as 'rescue' centres, as centres for research and education, as a primary means of discouraging unscrupulous collection in the wild—botanic gardens serve a multitude of purposes within the single overriding purpose of preventing plant extinctions.

David A. Munro
Director-General
IUCN

Contents

PART ONE

Rôles and Principles

PART TWO

National Policies and Activities

PART THREE

Education

PART FOUR

Background Support

PART FIVE

Special Groups

Introduction

J. P. M. BRENAN

Royal Botanic Gardens, Kew, England

The precursor to the present Conservation Conference was held from 2 to 6 September 1975, also at the Royal Botanic Gardens, Kew. The title of the 1975 Conference was 'The Function of Living Plant Collections in Conservation and in Conservation-Orientated Research and Public Education'. The Proceedings* of this Conference were published as the first of a NATO conference series on ecology (the Conference itself had been sponsored by the NATO Eco-Sciences Panel). That meeting was a notable success and achieved important results, not the least of which was a long series of agreed Resolutions (pp. 233-4) dealing with a variety of matters relating to the conservation of the world's plant life.

These Resolutions drew attention to such matters as the hazards affecting tropical areas, the need for a co-ordinated network both of reserves and gardens, the vulnerability of island floras and those of Mediterranean climates. They also focused attention on the needs of propagation, the value of seed banks, the ethics of collecting, the importance of publicity and also of knowing which species are in greatest need of conservation measures. In particular the Conference and the Resolutions it passed recognized clearly that the commitments of conservation are long-term, that the information necessary for sound co-ordinated conservation policies could not be assembled quickly or easily, and urged the desirability of continued study of certain important issues through the medium of working parties.

The 1975 Conference had as one of its principal objectives the exposure to those directing botanic gardens of the urgency of plant conservation and of the fact that botanic gardens have a genuine and important rôle to play in helping to solve some of the problems. I suspect that some directors and curators of botanic gardens found the latter concept a new and unfamiliar one. Quite shortly after the 1975 Conference the decision was taken that after a reasonable lapse of time it would be valuable to call another meeting to take the next step forward. As the title of the present Conference states, we are here looking at the practical rôle of botanic gardens in conservation. In doing this not only is the opinion of directors important but the advice and willingness of technical managers of living plant collections throughout the world. During the time since the last Conference man-made changes have adversely affected the world's flora at an

* SIMMONS, J. B., R. I. BEYER, P. E. BRANDHAM, G. Ll. LUCAS & V. T. H. PARRY (eds) (1976). *Conservation of Threatened Plants.* Plenum Press, New York and London. xvi+336 pp.

ever-increasing rate, and hence the need for the future help of botanic gardens
has become even more urgent.

In outline the main objectives of this Conference include the following:

1. To look at various aspects of the practical rôle that botanic gardens can play
   and to exchange information about them. Some of these aspects are indicated
   both in the titles of the sessions and of the individual contributions.
2. To see to what extent botanic gardens are now in a position to accept and
   fulfil practical commitments.
3. To establish policies for the future, to ensure that they are fulfilled, and to
   set up some means of co-ordinating and guiding future activities involving
   conservation in and by botanic gardens.

During the past three years it has been encouraging to see the awakening
interest of botanic gardens and others in their rôle as protectors of the international heritage of plants, although in some countries an awareness of the need
for plant conservation had taken root and developed earlier. The picture is now
a reasonably encouraging one and this is one of the most important aspects
brought out in this book.

The papers presented have been arranged by the editors into several more or
less natural sections. The first section bears the general title *Rôles and Principles*
and contains keynote papers that set out, through example, the practical rôle of
botanic gardens in conservation. The first three papers deal with threatened
plants in general, and so provide the essential background to later deliberations
on botanic gardens themselves. To begin with, Gren Lucas provides a brief
introduction to those activities of the major international organizations that affect
plant conservation and outlines the work of the IUCN Threatened Plants Committee, putting their botanic gardens projects in the perspective of their general
programme and the IUCN World Conservation Strategy. Sven Wahlberg highlights the threats to plants and vividly describes the Swedish Project Linnaeus,
a very successful example of a programme to conserve native plants and publicize
the need for conservation. He stresses the particular importance of involving in
such a project all sorts of people, often with no more than an amateur interest
in preserving the heritage around them, and shows how rare and beautiful plants
are an ideal peg on which to hang a general conservation message.

As various speakers repeatedly and rightly emphasize, conservation of plant
species in their natural habitat is almost always the most desirable policy.
However, in the United Kingdom, the rôle of the Nature Conservancy Council,
as explained by D. A. Ratcliffe, enables botanic gardens to concentrate on their
own special rôle in a way that does not always apply elsewhere. Even in the
United Kingdom, as he explains, valuable joint projects can be undertaken in
conjunction with botanic gardens; such a project, that of mapping the rare and
threatened species of eastern England, is outlined in a paper by S. M. Walters.
However, in other countries the administrative pattern may be quite different.
In a keynote paper on the rôle of a small and local botanic garden, David
Bramwell describes the enterprising work of the Jardín Botánico 'Viera y
Clavijo' in conserving many of the remarkable endemic species of the Canary
Islands, in some cases by stimulating the creation of small reserves and even
managing such areas themselves. At the same time he draws attention to the

need for international support for and encouragement to small, local botanic gardens trying to conserve large, diverse and highly threatened floras. The relationship of a botanic garden to a network of nature reserves is explored in a Polish context by B. A. Molski, who gives the exciting news of a new botanic garden and provides much valuable information on both threatened plants and protected areas in that country. The tropical scene is introduced by E. Soepadmo, who describes the formidable problems posed by intensive destruction of rain forest in South East Asia, outlines the probable massive losses of potentially useful plants and shows what is being attempted by botanic gardens, again with the good news of a garden set up relatively recently.

These considerations in a sense provide the framework essential for a botanic garden to pursue a successful policy in plant conservation, but how are the practical problems to be tackled? The basic needs when plants are to be conserved in a botanic garden are firstly to get the plants, secondly to establish and keep them and thirdly to distribute propagating material as widely as possible. All these aspects but in particular the first are dealt with in the paper by J. B. Simmons, the Curator of the Royal Botanic Gardens, Kew. A fourth task, to successfully re-introduce the species into its natural habitat, is vividly demonstrated by Louis Olivier's paper on re-introducing threatened sand-dune species into a locality on the Mediterranean coast of France.

If these papers provide the framework, the next section, under the title of *National Policies and Activities*, outlines the structure itself, giving as it does accounts of the conservation work of botanic gardens in a representative selection of countries around the world. The papers by E. E. Gogina and by A. V. Hall & H. B. Rycroft describe the enterprise with which plant conservation has been tackled in the USSR and in South Africa respectively, and the key rôle played in both countries by botanic gardens, not only in their own right but also in managing areas of natural vegetation that are of conservation importance. The work of the Threatened Plants Committee of the USSR Botanic Gardens Council, as described by Gogina, provides a conspicuous example of how virtually all the aims and rôles of botanic gardens in plant conservation, discussed in both the 1975 and the present meeting, are being tackled with success. Of especial interest is the production of a reference book that will act as a register of the rare and threatened species already in cultivation, an undertaking that echoes Gren Lucas' remarks on the TPC botanic gardens projects. To hear of the work of the 115 USSR botanic gardens in drawing up lists of threatened species, carrying out research on those plants, and making recommendations for new reserves and protected areas, in addition to all the more traditional tasks of botanic gardens, is very heartening. Some of the species involved are mentioned by B. A. Winterholler and the work of one particular garden, Tbilisi, by L.V. Asieshvili.

In some regions, however, botanic gardens face great difficulties and financial constraints, as is apparent from A. P. Vovides' account of the Xalapa Botanic Garden in Mexico. Here the problem has been tackled in a practical way and various surprising and unorthodox techniques have been used with conspicuous success, drawing great attention from the populace. Other difficulties may be biological, in particular the need to know how many plants are needed to make a 'survival-population' genetically viable in the long term and how this varies from species to species. The need for research on this crucial problem was

brought out by A. V. Hall and H. B. Rycroft from South Africa, a country with
a sadly large number of critically endangered species. They also explored the
potential of horticulturally restoring threatened species in the wild, a problem
about which very little is known at present but clearly one that has great
potential for the future.

Public opinion is a very important factor in almost all conservation work.
Public money is frequently involved, and often decisions have to be taken about
conflicting interests against strong economic arguments. Public appreciation can
be all-important. The need for 'selling' the idea of plant conservation to the
public is thus rightly and repeatedly emphasized by various speakers. Ratcliffe
touches it in his paper already mentioned. Sven Wahlberg vividly gives the
arguments from the viewpoint of the World Wildlife Fund in Sweden. In a
separate section on *Education*, Don Aldridge of the Countryside Commission
for Scotland amusingly illustrates how botanic gardens can attract the attention
of the public and so help in publicizing the needs and objectives of plant
conservation. That it is never too early to start is perhaps the message conveyed
by Gerhard Winkel's account of the School Biology Centre in Hanover. The
emphasis is on the children, but the parents are brought in too. How colour
television now enables the botanist to reach massive and hitherto unavailable
audiences is perhaps the parallel message of David Bellamy's striking film and
short talk. Even when the rôle of a botanic garden in plant conservation is an
accepted part of its policy, the continuing need for education and publicity is not
absent. As R. I. Beyer describes, ideas have continuously and repeatedly to be
conveyed to a changing staff, so that the enthusiasm and devotion which are so
important in the practical care of plants can be sustained year after year and
from generation of staff to generation.

Undoubtedly one of the most economic methods of holding living plants is in
the form of seeds in a seed bank. This aspect forms the main part of the fourth
section entitled *Background Support*. P. A. Thompson explains some of the
theoretical constraints and emphasizes the importance of preserving wild stands
in addition to cultivated ones on which emphasis has hitherto been placed; within
these constraints, however, a seed bank need not be too demanding in terms of
staff, space and facilities, as is indicated in the account given by Olaf Olsen and
Folmer Arnklit of the Seed Bank at Copenhagen Botanical Garden. The
encouragement of such banking facilities is one of the objectives of the OPTIMA
Commissions described by César Gómez-Campo.

Some groups are threatened more than others, as is evident from the final
section with its paper on orchids by Ingrid von Ramin, illustrating the difficulties
over cultivating the terrestrial species. Similar problems are discussed in an
earlier section by J. W. Wrigley, the Curator of the Canberra Botanic Gardens,
especially those of maintaining species in a comparatively harsh climate to which
they may not be physically suited. The collection of endemic orchids in Canberra
has posed special problems and evokes special solutions. So too does *Degenia
velebitica*, a threatened Crucifer of particular ecological and botanic importance,
on which has focused the attention of Ljubljana Botanical Garden as described
by Vinko Strgar.

It is a familiar problem in botanic gardens that rare species frequently start
from only one or a few original introductions. Especially when this is coupled

with substantial vegetative reproduction, the genetic range of such species may be artificially narrow in cultivation. P. C. de Jong of the Utrecht Botanic Garden describes how this problem is being dealt with in the Netherlands. The problems of maintaining a wide genetic stock, correctly labelled and protected from inter-hybridizing, are described by Sven Snogerup in relation to the endemic flora of the Aegean Sea; his paper clearly brings out the fact that practical problems alone make conserving species in botanic gardens a second best to protecting their wild habitats if there is any element of choice. This does not prevent, nor should it discourage, gardens from growing threatened species, but it highlights the need for them to be involved in field conservation as well. For species that are rare but not yet critically endangered, the botanic gardens stocks may be a second best for absolute conservation, but they may be of far greater value in educating the public, in drawing attention to the need for conservation, in providing material for research, and so on, all helping to ensure conservation of the site itself. It is this involvement with field conservation that was so dominant a theme of the Conference.

If we look at future policy for plant conservation and the part to be played by botanic gardens, certain points are clear:

1. The task of conserving the world's flora is far too great and diverse to be carried out by any one botanic garden, any one country or even in any one continent. Co-operation and co-ordination are essential.

2. The task is too big and too complex for a brief Conference to organize in detail; continuing subsequent effort is needed.

3. The present Conference was supported by the balance of funds from the 1975 Conference. The responsibility for continuing the work must be organized with other sources of help. One of the most important objectives of this Conference was to ensure that the guidelines for future organization are clearly indicated.

During the course of the Conference three Working Parties were set up to discuss various aspects of conservation in its relationship to botanic gardens. The appropriateness and importance of the subjects discussed need no further emphasis here. From the reports of the Working Parties present elsewhere in this book (pp. 8–14) it will, I hope, be seen that at least a good effort has been made towards the fulfilment of the third objective mentioned above.

The needs of conservation are worldwide and affect every country. Nevertheless it has been repeatedly and rightly pointed out that the vast majority of the world's plant life is within the tropics and that the magnificent climax ecosystem of tropical rain forest, outstanding in luxuriance and wealth of species, is perhaps the most vulnerable and endangered of all. It is here above all that future work, resources and assistance should be concentrated.

To all the delegates and participants who came from many countries—and often from great distances—to give their experience and knowledge in helping to solve problems and formulate policies, I extend my grateful thanks. I hope that the present volume will represent a substantial contribution towards meeting the needs of plant conservation and outlining the rôle that botanic gardens can play in this increasingly urgent and important matter.

CONFERENCE ON
THE PRACTICAL ROLE OF BOTANIC GARDENS IN THE CONSERVATION OF RARE AND THREATENED PLANTS

Agreed Conclusions

This Conference:

1. Agrees unanimously the following Resolution:

   Conscious that the rich tropical floras of the world are now in great hazard, this Conference firstly *urges* that a strong network of nature reserves and conservation-orientated gardens should be established throughout the tropics and subtropics, both through the strengthening and development of existing foundations and through the creation of new ones where the need exists; secondly *stresses* how essential it is that all countries where applicable should have a suitable network of national regional botanic gardens to fulfil their fundamental part in the management and conservation of natural resources, while recognizing that reserves are the basic focus for such schemes; and thirdly *urges* institutions throughout the world who are in a position to do so, to offer all possible help in this programme through technical aid, training and the secondment of personnel. To this third aim this Conference urges the setting up of Fellowships to provide a reciprocal interchange between botanic gardens for horticulturists and scientific staff in both the developed and developing regions, with an emphasis on conservation activities, this to be funded by the major multi-national and national corporations.

2. Calls upon the International Association of Botanic Gardens (I.A.B.G.) to clarify its rôle and enlarge its activities, being guided wherever possible by the suggested objectives outlined below:

   1. To promote educational programmes within botanic gardens, not only on themes of international plant conservation interest, but on all aspects of the work of botanic gardens and their associated herbaria;

   2. To encourage, and wherever possible assist in, the propagation of rare and threatened species, especially within reserves held by other organizations or by the botanic gardens themselves, and re-introduction work;

   3. To promote, and if possible organize, exchange of personnel between gardens, for training, for inclusion on local expeditions, and for increasing general awareness among botanic gardens staff of the holdings, capacities and operational problems and successes of other gardens;

   4. To co-ordinate on a continuing basis information on specialist holdings in botanic gardens, with the aim of reducing excessive duplication, of making best use of existing collections, and of stimulating special collections on a wide range of taxonomic groups both in seed banks and traditional cultivation;

6

5. To provide information on request about relevant scientific and technical developments, lists of species maintained in cultivation (where available from gardens), expedition programmes, collecting needs, and to receive from TPC information on which plants are rare and threatened in the wild;

6. To stimulate the formation of regional groups for the interchange of ideas and pooling of resources, e.g. seed banks;

7. To establish a system of awards relating to success in achieving these targets.

3. Identifies the urgent need for an organization to promote co-operation between botanic gardens on conservation matters. It invites the Threatened Plants Committee of IUCN to take on an additional commitment, to facilitate communication through an enlarged TPC Newsletter and to circulate lists of threatened plants among gardens to find out which species are in cultivation and where, and to publish the results. The Conference suggests that this programme be funded by small annual subscriptions from botanic gardens and promises full support to a group to be set up by the TPC entitled the 'Botanic Gardens Conservation Co-ordinating Body'.

4. Requests that both guidelines for collectors and leaflets to discourage casual collecting at home and abroad be formulated, agreed and distributed; that more consideration be given to the scientific aims of expeditions, to international legislation, to contact and co-operation with the host country's botanical institutions, to advance publicity on forthcoming expeditions, to sharing more widely the material and information gathered and to the discouragement of commercial collecting, as outlined in more detail in the Report from the Working Parties (p. 8).

Report from the Working Parties

Prepared by NIGEL TAYLOR and HUGH SYNGE

Three working parties were established on the first day of the conference. Introducing the three chairmen, Professor J. P. M. Brenan explained that one conference could not be expected to solve all the considerable problems of plant conservation, but that certain, particular aims could be discussed and valuable conclusions reached. This was the aim of the working parties approach. Each chairman was invited to briefly introduce the topic to be covered by his group and to outline its aims to the whole conference. Each Working Party held separate meetings during the conference and the various chairmen reported back to a plenary session towards the end, when the conclusions (pp. 6–7) were agreed.

The first group covered 'Codes of Practice for Collectors' and was chaired by J. B. Simmons. Introducing the subject he drew attention to the need for ethical guidelines for collectors, since there were always many expeditions in the field, and he went on to emphasize the importance of not dramatically reducing wild populations. He stressed the need for consultation with local botanists and suggested that once material was in a botanic garden it should be made available to other gardens throughout the world. Points made from the floor included the comment that botanists out of their home country tended to lack respect for the flora of the country they were visiting in comparison to their own flora. In particular, the representative of the Botanical Society of the British Isles, M. Briggs, commented that amateur botanists abroad tended to think they could validly collect all species that were not formally protected, and asked if a leaflet on collecting could be produced for distribution to travel agents and tour operators. From Sweden, S. Wahlberg mentioned his country's efforts to persuade travel bureaux to include in their advertisements a statement that manufactured souvenirs were preferable to picked flowers, under the memorable headline 'Corpses are not memories'. The point was made that a great deal of perseverance had been needed for tour operators to accept conservation principles, yet nevertheless holiday brochures were an ideal place to say "Enjoy the plants you will see but leave them where they are!"

Concern was also expressed about visiting expeditions neither giving adequate notice of their intentions to the local institutions and authorities, nor obtaining appropriate permission. Examples were given where one expedition arrived only to find another expedition with greater resources and knowledge had just set off into the field. Close co-operation was needed and it was suggested by A. Thompson that expeditions should give six months to one year's advanced notice to botanical organizations in the countries they intended to visit. At a later session A. R. Hassan King highlighted this point in a West African context and requested that countries who permitted collection of their native plants should

8

be informed as to what research and benefit was being derived from the plants that had been collected.

At the plenary session, a draft list of proposals on codes of practice for collectors was discussed and the following points were agreed:

1. Leaflets and posters should be prepared, published and distributed to Tour Leaders and Travel Agents, with the aim of discouraging uncontrolled casual collecting by members of the public, both at home and abroad, and with a warning of possible penalties.

2. Practical guidelines for collectors must be formulated, agreed and distributed.

3. To be acceptable any expedition must have valid scientific aims. These were identified as:

   (a) Conservation
   (b) Research
   (c) Education
   (d) Exploration.

4. Expeditions must make contact well in advance with local botanical institutions and study regional, national and international regulations, in particular the provisions of the Convention on International Trade in Endangered Species of Wild Fauna and Flora. They must also obtain appropriate permission to collect, and seek mutual co-operation and collaboration. Wherever possible collecting data with a duplicate set of material collected should be deposited in the host country, who should also receive copies of any resulting publications.

Some consideration was given to a further proposal on recording all expeditions in which botanic gardens, universities and other scientific institutions were involved. However desirable and valuable such a scheme would be, it was generally felt to be impractical and unlikely to succeed. In particular J. P. M. Brenan emphasized the difficulties in obtaining information on the frequently self-financed expeditions of university students, etc., and G. Ll. Lucas remarked that lists based on organizations volunteering information on their proposed trips would pick up only a small percentage of the actual number in the field and then only if prepared on a national basis. Nevertheless the point was firmly made that there was a great need for collaboration and co-operation at all levels, especially between botanists on visiting expeditions and botanists in the host country.

Discussion also centred around responsibilities for collecting and bulking up endangered species in gardens. It may be remembered that at the previous conference in 1975 the point had been emphasized again and again that involvement in conservation was not a mandate to collect material of threatened species; rather, one should first ascertain whether the species were in cultivation elsewhere. If not, then the botanic garden of the country concerned should first be requested to gather limited material, preferably as seed, for bulking up and then distribution to other botanic gardens and institutions. Only if such gardens were unable to do this should private collectors or collectors from foreign botanic gardens gather material of these species, and then only with the appropriate governmental approval and support. In the discussion at this meeting, the need for bulking up such species in the countries of origin was again emphasized; this could be particularly important for plants such as orchids and cacti which are threatened by collecting and trade. There was a clear market for well grown material, and through the Convention on International Trade in Endangered Species of Wild Fauna and Flora there was also a means of controlling exports (and imports) of wild-collected as opposed to propagated stocks. Growing these plants on a commercial scale for the horticultural market could provide valuable income and foreign exchange for countries rich in these desirable but declining plants.

The second group, on 'Responsibilities for Holding Non-European Endangered Endemics' was chaired by D. Bramwell, who, in an introduction on the work of this group, said:

"I am sure all of us here are convinced by the basic arguments for plant conservation. We appreciate that the ideal solution is conservation in the field, in natural habitats; but the worldwide problem of destruction of natural vegetation and extinction of plant species is such a serious one that we cannot wait for that utopian situation where all countries have adequate conservation legislation and the goodwill and economic means to enforce it. We all know that by the time Utopia arrives it will simply be too late.

"Botanic gardens, therefore, as the holders of living plant collections, must have an important rôle to play in ensuring the immediate survival of endangered species. Europe and North America have a relatively large number of botanic gardens covering a wide range of habitats and also have enough trained staff to handle most of the European and North American endangered species if a practical, co-ordinated policy is established. The situation in tropical and subtropical regions outside Europe is, however, rather different. In very few areas are there active centres capable of making a major contribution to local conservation. In many places gardens exist which potentially could be involved in conservation but, in the poorer areas of the world where minimal financial resources are available, priority will not be given to conservation projects. As R. L. Shaw pointed out at the 1975 conference, 'The scope of some botanic gardens is severely curtailed for political and economic reasons.'

"We are, when talking about conservation in gardens, I hope talking of the conservation of representative gene pools for practical research and for rehabilitation in natural ecosystem reserves and not just as museum collections. As J. Heslop-Harrison said in 1975, 'We have to consider the commitment to be an open-ended one'.

"In order to achieve success outside Europe, North America and the USSR, etc., the rich nations must take on some, perhaps initially a major part, of the financial burden for conservation. I am sure we could in a number of key areas establish regional centres for holding endangered species, but not just for holding them: for basic research on breeding systems, reproductive biology and so on to ensure their survival in the garden and in the field and for their eventual rehabilitation when possible.

"The key to the whole business of conservation in the under-developed parts of the world is, of course, finance—and I believe we must make a start and try to raise the money necessary. At the previous conservation symposium in 1975 the then Director-General of IUCN said 'If botanical gardens are engaged in conservation work, they would receive support from the international programme'. Now to my naïve and simple mind 'support' can only mean financial support; more words will not help.

"Many gardens in Europe do a magnificent job in maintaining endangered species from other regions of the world, but I believe that a real effort must now be made to get the necessary finance from the international agencies or elsewhere to transfer this burden back to the regions involved—in the long run it will be cheaper and more effective, and if sites and centres are well chosen then the developed world will have a major rôle to play in providing the technical aid and the know-how which is lacking in under-developed regions. Our responsibility for non-European endangered endemics is not just to grow a few of them, but to find the means by which local centres in the areas involved can do the job with our help and full support.

"The proposed botanic gardens conservation organization should, perhaps, be prepared to take on the rôle of fund-raiser and provider of technical aid for future development."

In discussion, the following points were made:

That however much support local botanic gardens received from organizations such as IUCN, there would still be insufficient money to completely ensure security for endangered species in the under-developed world; therefore active gardens should take on the cultivation of threatened plants from similar climates in addition to growing their own endangered species. This had been done by botanic gardens in Hawaii (J. P. M. Brenan).

Since financial aid was at the nub of the problem and since the organizations that fund botanic gardens in the developed world were either absent or unwilling to assist in under-developed countries, it was government support that was most needed (C. D. Adams).

That emphasis should be given to unstable ecosystems such as those of the Mediterranean basin since these had already been most affected by man and most needed the resources and co-operation of scientifically advanced countries in training or even in the provision of staff (M. Avishai).

That a centre for training staff in wildlife conservation was required (A. R. Hassan King).

Turning to the practical problems of cultivating threatened plants, J. P. M. Brenan emphasized that if botanic gardens were attempting to ensure the survival of species in cultivation, it was essential that genetically representative samples were grown. Ideally, examples from the full range of a species' variation should be maintained. Concern was expressed by H. B. Rycroft at the risk of selection of 'improved' forms in botanic gardens; D. Bramwell suggested the value of seed banking in this respect.

This Working Party then presented at the plenary session a draft resolution, which was accepted with minor amendments as the first section of the Agreed Conclusions of the Conference (p. 6).

The subject of the third group, 'Communications—the Need for a European Co-ordinating Body and Its Practical Objectives', was introduced by its

Chairman, S. M. Walters, who reviewed the period since the 1975 conference, drawing attention to the improvement in communications that had occurred between botanic gardens since that date. He pointed out how information on which plants were rare or threatened in the wild was now forthcoming from the IUCN Threatened Plants Committee, and that this information was essential to strengthen a botanic garden's conservation activities. Also important was to know what other gardens were doing and here the need was for a newsletter or journal.

After discussion it was agreed that the working party should not just restrict itself to a European co-ordinating body but should consider an organization (or organizations) to cover all botanic gardens.

Two short papers were presented to the Working Party: one, entitled, 'Proposal for the formation of an I.U.B.S. Commission', proposed that a Commission of the International Union for Biological Sciences be created to act as an international watchdog on plant conservation, making representations at Government level to advise where urgent action was required, and promoting plant conservation in general, as well as acting as a co-ordinating body for the conservation activities of botanic gardens.

The second paper, entitled 'Suggestions by IUCN Threatened Plants Committee Secretariat on the proposed new organization', suggested seven objectives as a starting point for discussion. (These were subsequently taken up as section 2.1–2.7 of the Agreed Conclusions.) The objectives all related to botanic gardens' work in conservation. The paper also briefly reported on the existing TPC project to find which species known to be rare or threatened in Europe were already in botanic gardens, and the need to expand this work throughout the world. Attention was drawn to the fact that one of the three arms of the TPC, as envisaged by its Chairman, J. Heslop-Harrison, was an institutional network of collaborating botanic gardens.

These documents were very fully discussed in the Working Group meeting and many valuable points made, relating to the organization of international conservation on the one hand (IUCN in particular) and the collaborative mechanisms between botanic gardens (I.A.B.G.) on the other. Many members stressed the importance of using existing structures and not proliferating new organizations. Nevertheless it was widely regretted that I.A.B.G. did not at present have the facilities, in particular financial, to undertake what was seen as essential activities in furthering collaboration between botanic gardens, and developing their conservation rôle.

The conclusions the Working Party reached were as follows:

1. There was little support for a Commission on Conservation in I.U.B.S., because of overlap with IUCN; nor was there much support for a formal European section of I.A.B.G., despite the evident success of the American section (A.A.B.G.), since the communications problem concerned more the botanic gardens outside Europe; communication and collaboration within the continent was already enhanced by the existing TPC European Botanic Gardens Conservation Project and through the *Gärtnerisch-Botanischer Brief* which fulfilled the rôle of a European botanic gardens journal.

2. Rather, the Working Party favoured a renewed approach to I.A.B.G. indicating

the evident desire of botanic gardens for effective action, both in those areas of development not directly concerned with conservation and in the more conservation-orientated aims listed in the TPC Secretariat's seven points.

3. The group also expressed appreciation at the existing TPC services* and requested a modest expansion, especially in the circulating of lists of threatened plants to ascertain which species are in cultivation, and in the distribution of its Newsletter.

4. An offer by B. A. Molski for a new international journal on botanic garden conservation was welcomed for further exploration.

Originally a fourth working party had been proposed under the title 'Provision of Guidelines for Governments (Regional and Central) with Regard to the Botanic Garden Rôle in Conservation'. The Chairman designate, G. Ll. Lucas, told delegates that on the European list of rare and threatened plants† was a resolution for governments listing plant conservation recommendations; one of these was adequate support both practical and financial for botanic gardens. The document had been signed by the Committee of Ministers (of the Council of Europe) and is given as Appendix 1, with appropriate additional comment. The Conference was also reminded of the 1975 conference resolution, which is given as Appendix 2. Thus it was felt that a fourth working party was not necessary and any additional recommendations and guidelines for governments that were needed should be considered by the second working party, that on 'Responsibilities for Holding Non-European Endangered Endemics'.

After the three working parties had reported at the plenary session, the Chairman, J. P. M. Brenan, invited G. Ll. Lucas as Secretary of the IUCN Threatened Plants Committee to reply to the comments about an expansion of TPC services, made by S. M. Walters, and to suggest what could be done by TPC about collaboration between botanic gardens. Mr Lucas responded by saying:

"We have all given a lot of thought to the problems raised in maintaining effective action and collaboration among botanic gardens on conservation and to the fact that above all continuing activity is essential. The difficulty for the TPC, however, is that nothing succeeds like success. The European project for botanic gardens is running smoothly and you have all been very complimentary in your comments and suggestions that we should now take on the world!

"However, the third arm of TPC activity as originally envisaged was to co-ordinate botanic gardens in their conservation aims, but until now we had believed that other bodies were involved and better placed to carry out this work. It now seems that this is not the case and having talked to quite a number of the delegates, I would like to suggest the following idea: that the Threatened Plants Committee of IUCN sets up a 'Botanic Gardens Conservation Co-ordinating Body' whose first task would be to take on the existing TPC European project and bring it to a conclusion, producing a final report indicating which threatened species are in cultivation and where. Due to severe limitations of staff and

---

* Described in detail in Gren Lucas's paper (pp. 15–23). In particular this refers to the TPC European Botanic Gardens Conservation Project.

† IUCN Threatened Plants Committee (1977). *Rare, threatened and endemic plants in Europe*. Nature & Environment Series No. 14, Council of Europe, Strasbourg.

money, the TPC Secretariat itself cannot yet go further than its present objective of identifying the threatened and endangered species of the world, and thus it would only be able to provide the Co-ordinating Body with lists of threatened plants and a focus for particular problems. However it would be foolish not to go one step further and make use of the data. Thus I would suggest a separate Secretary and Treasurer would need to be appointed under the Threatened Plants Committee. I have already talked to some of you and tentatively suggested a small sub-committee to oversee the work—one does not want a large formal structure; the success of the TPC is that you all generously help when you can. David Bramwell has kindly agreed to act as Secretary, and, with the Curator's and Director's permission, Ian Beyer will act as Treasurer.

"Regretfully the TPC does not have a budget for the next two or three years that can cope with this extra work. Therefore the scheme will have to be self-financing but it will only be the bare essentials. This will cover the circulation of the lists of threatened plants, the collation and publication of the resulting data as to which of those species you have in your gardens, and the distribution of an expanded TPC Newsletter to keep you in touch with what we are doing. It will also cover the costs of maintaining the correspondence of the Sub-Committee. I hope you will *all* be part of this Body and that it will be the solution to your ideas and needs".

Later on G. Ll. Lucas emphasized that where there were existing bodies co-ordinating national or regional activity, such as the Threatened Plants Committee of the USSR Botanic Gardens Council (see pp. 141-6), then the TPC Body would co-operate with them and would in no way duplicate their work.

After discussion the full proposal was agreed and accepted by the conference. It was agreed at the same time that the I.A.B.G. should be called upon to clarify its rôle and enlarge its activities, being guided wherever possible by the seven suggested objectives outlined during the meeting. This statement is given in full in the Agreed Conclusions (pp. 6-7).

At the close of the Proceedings, S. Wahlberg provided a timely reminder when he said:

"We are very satisfied with ourselves just now, having put together this resolution. We will now go home and many will do nothing, but it is when we go home that the real work should begin. For instance, how much have the European delegates in this room done in furthering the resolution on the conservation of rare and threatened plants in Europe made by the Committee of Ministers? How many have been to the Minister in their own country and asked 'Mr Minister, what have you done on your own resolution'? You may say to him: 'Sir, you now have six months. After that I will talk with a member of the opposition and he will ask you difficult questions in Parliament'. I propose that you think about this afterwards and think who is the right man for getting the points of this resolution through in your own country. I think that you should read it through again and go to a member of Parliament or a Minister and also confide in whoever is the actual man doing the work within the appropriate Ministry. Perhaps you are not the right person to approach him, but then ask someone else to see him or go to his boss. I think this is the most important part of the conference and the part which we always tend to forget."

PART ONE

Rôles and Principles

Organizations and Contacts for Conservation Throughout the World

GREN LUCAS

The Herbarium, Royal Botanic Gardens, Kew, England

With the growing awareness that the plant kingdom is our life-support system, an ever-increasing number of people are asking what are the threats to plants, what consequences will result from overall losses and degradation, what is being done to protect plant habitats, and how should they, as individuals and members of decision-making organizations, be contributing to plant conservation. To answer these questions one needs data; many of those who manage or curate European botanic gardens will be familiar with the work of the IUCN Threatened Plants Committee and the data-gathering work it is carrying out, in particular with its botanic gardens project set up as a result of the previous conservation conference held at Kew in 1975. In this paper I want not only to give the preliminary results of this work, but perhaps more importantly, to show how it is a part, albeit a small one, of a general movement throughout the world, national and international, aimed towards the conservation of the plant kingdom. In particular I want to highlight how it is an essential complement to the general task of TPC in identifying which plant species are rare and threatened in their wild habitats. Slowly but surely, with your support and that of all our other colleagues, we are identifying these species, where they still occur, and what needs to be done to ensure their survival. Equally vital, and especially so in this context, is to know who has them in cultivation.

INTERNATIONAL ORGANIZATIONS

Firstly let me outline, very briefly and in a very personal light, how I see the rôles of the major international organizations that are involved with conservation, in particular, how we can work together to our mutual advantage, even to a stage where they may be willing to provide financial support for some of our work in a conservation context.

15

FAO.   The Food and Agriculture Organization of the UN is perhaps the most prominent international body that is concerned directly with plants. Its research programmes in conservation of crop species and their progenitors are an admirable example of what can be done. So too are the seed banks for crop cultivars and their allies linked together under FAO. The work of Sir Otto Frankel and his collaborators on the theory, practice and difficulties, both biological and practical, of conserving crop genetic resources for the future benefit of mankind is of the greatest importance and has shed light on many aspects of conservation relevant to non-crop species (Frankel & Bennett, 1970; Frankel & Hawkes, 1975). But there are many more plants that we should protect; the contrast between the estimated 25,000 species dangerously rare and severely declining towards extinction, and the less than 50 species of crop plants covered by FAO, is most striking and brings the problem faced by other organizations sharply into focus.

A whole area that has been neglected hitherto is the conservation of species of horticultural value, both in terms of protecting the species in the wild and maintaining the range of cultivars and other variants in gardens. It is, however, encouraging to report that in England the Royal Horticultural Society has taken a lead and in October 1978 arranged a most successful symposium on the problem (Brickell, 1979). A vital outcome of this meeting is to find out which cultivars of the more old-fashioned garden plants are becoming rare—one immediately thinks of carnations (*Dianthus*), sweet peas (*Lathyrus odoratus*) and old roses. An encouraging development being actively discussed is for a body like the RHS to take on the responsibility of growing commercially these species and important cultivars whose continued production nurserymen find uneconomic. In particular it is hoped that some of the temperate trees and shrubs until now propagated by Hillier's world-famous nursery will continue to be available and distributed to ensure long-term survival of cultivars in this way.

WHO (The World Health Organization).   In recent years, the spotlight of the medical world has tended to turn away from the highly sophisticated synthetic drugs, back to those natural medicines that are produced from plants and have been used in tropical countries for generations. The WHO has been at the forefront of this movement and is involved with research programmes to locate and study these crucially important old drug plants. A vital input to this work is the information on local plant uses that can be obtained from the label data on many herbarium specimens. There is a great need to expand the production of works of reference based on this data and on allied sources. Another way botanic gardens can help is by maintaining, propagating and supplying material of the individual species under study. Already many gardens are acting as primary or third party quarantine stations, as does the Royal Botanic Gardens, Kew, with the aim of allowing new and potentially useful crops to be taken all over the world without endangering existing crops through introduced pests and diseases. It seems WHO are not yet aware of the immense conservation contribution botanic gardens could make in this field. The Cambridge Botanic Garden's involvement with supplying garden-propagated plant material under contract to ICI shows just what can be done (see p. 45).

THE WORLD BANK.   As conservation is probably best defined as the wise use of natural resources in the long term, ecology and conservation must be an essential part of the brief of organizations like the World Bank that are concerned with developing and making use of natural resources for the long-term benefit of mankind. It is being increasingly realized that the future potential of the developing world, where the Bank is particularly active, will depend not upon the introduction of heavy industry but upon the controlled use of those natural resources of plants and animals on which such countries have always depended in the past. As managers of plant collections and as individuals concerned about plant conservation, we should make greater efforts to become involved with the massive programmes and projects funded by the World Bank. The effect of such projects on the environment can be staggering; however, ecological impact assessments are now a required part of their planning and these statements should include data on threatened species, the very data that we may be able to provide. We must certainly encourage their awareness.

One particular priority is for the development of horticultural crops to pay for local conservation. This is still a sadly under-used approach. In Brazil, however, such a programme has already started: wild orchids are gathered from areas whose vegetation is to be, or has been, devastated, the orchids are propagated and the proceeds of their sale contributes to the National Park budget. A longer term strategy will be for the inclusion of propagation centres in or close to National Parks. Their task will be to bulk up and sell horticulturally desirable plants such as orchids, bromeliads and cacti that may occur within the Park boundaries; this would relieve pressure on wild populations, in particular reducing illegal poaching inside the Park itself, generate local income through employment, and provide funds for the National Park Budget.

UNEP.   The United Nations Environment Programme was set up after the highly successful UN Conference on the Environment at Stockholm in 1972. Under the energetic leadership of Maurice Strong, it has developed the key rôle of helping and persuading governments to elaborate strategies which make conservation planning an integral part of development. Its two most noted successes are the stimulus it has provided towards cleaning up the Mediterranean Sea, through a series of major conferences and protocols in which virtually all the Mediterranean nations have amicably participated, and secondly the attention it has drawn to the tragic problem of desertification, in particular through the UN Conference on Desertification in 1977, and through a whole range of ancillary projects that it has supported.

IUCN.   Founded in 1948 by Sir Julian Huxley while Director-General of Unesco, the International Union for Conservation of Nature and Natural Resources is the main international body concerned primarily with nature conservation. It has a remarkable constitution in that the membership is made up of sovereign states, of government bodies, of independent national and international conservation bodies and in a few cases of individuals. It is not part of the UN family and so has the advantage of being able to offer advice to states without having to be asked to do so first. The Union meets every three years at a General Assembly when its programme is set for the following session. Day to day activities are managed by the Secretariat in Morges, Switzerland, and through the six Commissions, of which one, the Survival Service Commission under the Chairmanship of Sir Peter Scott, covers endangered species. It is under this Commission that the Threatened Plants Committee operates.

At present IUCN is developing a World Conservation Strategy, with the financial assistance of UNEP and the World Wildlife Fund. The section on plants and the detailed recommendations on action for threatened plants were put together by the Threatened Plants Committee Secretariat, based upon the data delegates and colleagues all over the world have supplied. The aim of the Strategy is to show exactly what are the fundamental requirements for conservation, in terms of a government's policy, of public education, of scientific research and so on. It will then suggest how these requirements may best be met and attempt to provide a clear but concise explanation as to why these requirements should be met—in other words why conservation is essential to all our futures. The detailed recommendations that follow from this outline, in the words of IUCN, focus also on "the ecosystems and species of such importance and facing such grave problems that they require urgent measures either before or at the same time as the fundamental requirements are being met". The essence of the Strategy, highlighted throughout the draft discussion document, is the need to promote conservation as a part of development. Conservation cannot succeed in isolation.

All these organizations, large as they are, cannot function without the data and the expertise that can only come from the field workers. It is their knowledge, and the manpower of those numerous individuals and establishments all over the world that enable these organizations to succeed in larger strategies.

Thus it is particularly alarming to see the current lack of data on threatened plants, especially for the tropics, and to realize that so much of the data we do have comes from unofficial, often amateur botanists. The provision of this essential information, area by area, species by species, cannot be the principal task of experts like yourselves; it has to be a sideline to your main research and curatorial responsibilities. Nevertheless the input that you can make is absolutely crucial and fundamental to the success of the whole approach. It was with this in mind that the Threatened Plants Committee of IUCN was formed, the aim being to provide a clearing-house for information on threatened plants, and to stimulate the gathering of this data from areas where the process was not already underway.

THE THREATENED PLANTS COMMITTEE

The TPC receives funding from the sister organization of IUCN, the World Wildlife Fund, and without this no progress could have been made. Currently this pays for a small Secretariat, housed in the Herbarium at Kew and consisting of one Research Officer (Mr H. Synge), one part-time typist and use of a computer facility. It is hoped that money will be forthcoming for other posts during 1979 as at present TPC is grossly overstretched and cannot fulfil its current tasks without further manpower.

The backbone of the TPC is a network of over 250 correspondents from all over the world; most of them are key botanists in their own countries. All are kept in touch with TPC activities through a twice-yearly Newsletter, as well as by our continual requests for information and for help on specific problems.

The three major spearheads for TPC activity were outlined by the chairman at the very beginning (Heslop-Harrison, 1974). These are:

(a) Regional Groups.
(b) Taxonomic Groups (e.g. palms, cycads).
(c) Institutional Groupings (in particular of botanic gardens).

How does this international conservation co-operation through the TPC work? How can one country help another country in this field? It is usually the same problems that need solving and often the same species needing protection. In the four years of the existence of TPC, progress has been made on all the three fronts listed above, but to answer these questions let me briefly outline one of our regional activities, that for Europe. I know better than most that this is the easiest area to cover, due to the density of botanists available and due to the limited size of the flora, which is itself relatively well known and uniformly covered by *Flora Europaea* (Tutin *et al.*, 1964–79). Having screened all existing national lists of threatened plants and then distilled from the Flora a first list of candidate species, we circulated country lists of these plants to all who even hinted that they were interested in conservation or in the field status of the plants. This gave rise to the draft list of threatened species, the individual species being annotated with the IUCN *Red Data Book* categories to indicate the degree of threat to each. Many problems remain to be tidied up but this refining and updating is a continual process. However, a draft list was presented to the Council of Europe (who had in part commissioned the original work) and this included detailed recommendations for action. These recommendations, repro-duced as Appendix 1, have been endorsed by the governments of the Council of Europe and now provide a powerful lever for conservation bodies in each country. The final list has now been accepted by these governments and published by the Council of Europe (IUCN Threatened Plants Committee, 1977).

Furthermore, a list of 102 *Endangered* European species prepared by the TPC in October 1978 will form the Appendix of plants in the Council's European Wildlife Convention; under the terms of the present draft, now rapidly being completed, ratifying nations will be required both to give these species full legal protection against picking and uprooting, and also to protect their habitats in appropriate reserves.

The European activities illustrate well one of the tenets of TPC activity. Firstly botanists and others on the ground provide the data, bit by bit. This is then assembled and used by international and national organizations, who between them are able to create the climate of opinion that enables those active on the ground and at local level to protect the individual species identified as threatened, or to use the data to help promote the protection of particular habitats. The Council of Europe's resolutions and conventions provide a stimulus for action by governments in supporting the detailed conservation work itself. Our aim is that no more plant species shall become extinct in Europe and that all shall be protected in reserves; judging by the flurry of national activities now underway, this is not an unrealistic goal. It is these activities rather than the List itself that are the sign of successful conservation.

We realize, however, that progress will be much more difficult in other regions, though experience so far has been encouraging. The TPC listings for North Africa and the Middle East (in collaboration with OPTIMA—see pp. 195-7) are nearing a draft stage, those for the Caribbean islands are well underway, and a number of national lists have been received for tropical Africa, the next main region to be covered, here in collaboration with the Association pour l'Etude Taxonomique de la Flore d'Afrique Tropicale (AETFAT). We hope to start a major project on Central and South America during 1979, with emphasis on centres of endemism, and a list of threatened plants for the Pacific will be next in line. Programmes for South East and Western Asia are already in sight. These projects aim to complement the existing lists for the USA (Ayensu & DeFilipps, 1978), the preliminary data being elaborated at present for Canada (Argus & White, 1977, 1978; Maher *et al.*, 1978), the list for Southern Africa (Hall, de Winter & de Winter, in press), the initial state lists for Australia now in need of refining and updating (Specht *et al.*, 1974), the data sheets for New Zealand (Given, 1976-7) and the Red Data Books for the USSR (Borodin *et al.*, 1978; Takhtajan, 1975—see also E. E. Gogina's paper, pp. 141-7). The aim is that by the next IUCN General Assembly in the autumn of 1981 there will be at least an initial statement on plant conservation for virtually all countries of the world. But these lists and reports, it must be remembered, are just the beginning.

Another result of the information we have received has been the production of *The IUCN Plant Red Data Book* (Lucas & Synge, 1978), which gives detailed case histories on 250 threatened plants carefully selected from all parts of the world, to indicate the threats and the types of plant affected, as well as to highlight those threatened species of potential value to man, be it economic, horticultural or medicinal. More important, we hope it will draw attention to the growing and continuing threats to the world's natural ecosystems and the diversity of plants they contain. The intention was to provide a basic statement on plant conservation, with examples to suit every need, so being an essential source of information for science writers, journalists and teachers, as well as those in a direct position to protect the species included as threatened. All this data comes from your individual activity without which TPC would be nothing.

Botanic gardens are poised to play a very crucial rôle in plant conservation, and one of the recommendations in the Council of Europe Ministers' Resolution is that botanic gardens should receive appropriate governmental support for them to be able to carry out this rôle. As is shown so well by the papers presented here, this rôle goes far beyond simply growing the threatened species in the garden as a long-stop against final extinction in the wild. It includes, for example, holding small species reserves and undertaking the type of horticultural work outlined by H. B. Rycroft & A. V. Hall in this book. It includes helping conservation bodies identify which are the key sites to be protected and in aiding the habitat management of these sites. It includes the re-introduction of locally extinct species following bulking up in the garden. It also includes using the material for education and for supplying research needs. It was with all these tasks in mind that TPC launched a project to find out which European threatened species were in cultivation and where.

THE EUROPEAN BOTANIC GARDENS CONSERVATION PROJECT

At the previous Kew Conservation Conference in 1975, it was agreed a project be launched to look into the holdings of threatened plants in botanic gardens. The Chairman of the European Sub-Committee of the TPC, S. M. Walters, prepared an introductory paper which was circulated to all gardens in Europe (Walters, 1977). This paper explained the rôles of botanic gardens in conservation, as envisaged by the conference, and asked those wishing to collaborate in an initial data-handling scheme to return a short form to the TPC at Kew. The TPC would then circulate lists of threatened plants to each garden who had responded and ask them to mark off which species they had in cultivation. Replies were forthcoming from 109 establishments in Europe and all received in return a copy of the IUCN list of threatened and endemic plants for their country. These lists, annotated to indicate which species were in cultivation and which of these were of known wild source origin, have been returned from more than half of these gardens. Subsequently, gardens were sent the full European list for annotation; this contains 1878 rare and threatened taxa. So far 481 species have been shown to be in cultivation and the number is increasing each week as the replies come in. We hope to publish an initial report on the project during 1979.

Already some general points are emerging:

1. It is the horticulturally well-known and attractive species that tend to be the ones in gardens, with, for example, high percentages in the Liliaceae, Primulaceae, Campanulaceae, Ranunculaceae and Saxifragaceae. It is clear that efforts must be made to ensure the less attractive threatened species are also cultivated.

2. Of the species so far recorded 213 are from one garden only. We must try and spread the load.

3. The five most abundant threatened species in gardens are *Salvinia natans* (17), *Cypripedium calceolus* (20), *Dianthus gratianopolitanus* (21), *Marsilea quadrifolia* (25) and *Pilularia globulifera* (28). All these are widespread species, but ones very rare or seriously threatened throughout their European range, either by collecting (as in the case of *Cypripedium*) or by destruction, drainage or general degradation of their wetland habitat (as in the case of *Salvinia* and *Marsilea*).

This work, it is hoped, can provide the basis for helping gardens in the choice of what to grow and from which garden material of particular threatened species may be obtained. It must not be a charter to devastate the wild populations; instead, as outlined in the report from the Working Parties (p. 9), if a threatened species is almost certainly not in cultivation, one botanic garden in the country in which it grows should first be asked to gather limited material for bulking up and eventual distribution.

Once the basic data has been put together, it is relatively simple to keep it up-to-date, provided all continue to co-operate, replying to routine enquiries once or twice a year. How much more can our normal activities benefit the plant kingdom and mankind with very little extra work on our part? International bodies in the field of conservation are only successful if they receive help and information at the local level. This has been conspicuously proved in the case of the Threatened Plants Committee and I would like in conclusion to take this opportunity of thanking publicly all those who have contributed help and information. Success in your jobs means success in international conservation.

REFERENCES

AYENSU, E. S. & R. A. DEFILIPPS (1978). *Endangered and Threatened Plants of the United States.* Smithsonian Institution and World Wildlife Fund, Inc., Washington, D.C. xv+403 pp.

ARGUS, G. W. & D. J. WHITE (1977). The Rare Vascular Plants of Ontario. *Syllogeus*, **No. 14.** 63 pp.

——(1978). The Rare Vascular Plants of Alberta. *Syllogeus*, **No. 17.** 46 pp.

BORODIN, A.M. *et al.* (eds) (1978). *Red Data Book of USSR.* Lesnaya Promyshlennost, Moscow. 459 pp. (In Russian).

BRICKELL, C. D. (1979). The RHS Conservation Conference. *Garden* (UK) **104(4):** 161–171.

FRANKEL, O. H. & E. BENNETT (eds) (1970). *Genetic Resources in Plants—Their Exploration and Conservation.* IBP Handbook No. 11. Blackwell Scientific Publications, Oxford. xxi+554 pp.

FRANKEL, O. H. & J. G. HAWKES (eds) (1975). *Crop genetic resources for today and tomorrow.* International Biological Programme 2. Cambridge University Press. xix+492 pp.

GIVEN, D. R. (1976, 1977). *Threatened Plants of New Zealand.* Botany Division, DSIR, New Zealand. Loose-leaf, mimeo.

HALL, A. V., M. de WINTER & B. de WINTER (in press). *Threatened and rare plants of Southern Africa.* South African National Programmes Report. Council for Scientific and Industrial Research, Pretoria.

HESLOP-HARRISON, J. (1974). Postscript: The Threatened Plants Committee. *In 'Succulents in Peril'* (ed. D. R. Hunt). Suppl. to *Bull. Int. Org. Succ. Pl. Study* 3(3). Pp. 30–32.

IUCN THREATENED PLANTS COMMITTEE (1977). *List of rare, threatened and endemic plants in Europe.* Nature and Environment Series No. 14, Council of Europe, Strasbourg. iv+286 pp.

IUCN (in press). *The World Conservation Strategy.* IUCN, Morges, Switzerland.

LUCAS, G. LL. & A. H. M. SYNGE (1977). The IUCN Threatened Plants Committee and Its Work Throughout the World. *Envir. Conserv.* **4(3):** 179–187.

——(1978). *The IUCN Plant Red Data Book.* IUCN, Morges, Switzerland. 540 pp. (Available from TPC, c/o Herbarium, Royal Botanic Gardens, Kew).

——(1978). Higher Threatened Plants. *In 'A Sourcebook for the World Conservation Strategy'.* Manuscript circulated at the 14th IUCN General Assembly, Ashkhabad, USSR, October 1978.

MAHER, R. V., D. J. WHITE, G. W. ARGUS & P. A. KEDDY (1978). The Rare Vascular Plants of Nova Scotia. *Syllogeus*, **No. 18.** 37 pp.

SPECHT, R. L. *et al.* (eds) (1974). Conservation of Major Plant Communities in Australia and Papua New Guinea. *Austral. J. Bot., Suppl. Ser.*, **7.** 667 pp.

TAKHTAJAN, A. L. (ed.) (1975). *Red Book: Native Plant Species to be Protected in the USSR*. Leningrad. 204 pp. (In Russian).

TUTIN, T. G. *et al.* (eds) (1964–79). *Flora Europaea*. Cambridge University Press. 5 vols.

WALTERS, S. M. (1977). The rôle of European Botanic Gardens in the Conservation of rare and threatened plant species. *Gärtn.-Bot. Brief* **51:** 2–22.

God Created, Linnaeus Arranged
Project Linnaeus, an Effort to Save that Good Work for the Future

SVEN WAHLBERG

World Wildlife Fund, Sweden

"At this season Nature wore her most cheerful and delightful aspect, and Flora celebrated her nuptials with Phoebus."

Linnaeus, Lachesis Lapponica
Translated Troilius & Smith
London, 1811

The last ten years and more have given me ample possibilities to travel all around the world. The more I see the more I learn, and the more I am convinced that the cutting of the forests and the destruction of the flora in the warmer parts of the world is the greatest threat to all life on earth. We have all read what was written in the old times. Virgil told us how lush the Mediterranean landscape was, a landscape which now must be considered a desert or a semi-desert. The woods are cut down, the slopes are eroded and we have all heard Marlene Dietrich singing "Where have all the flowers gone?"

In Bukhara in Uzbekistan, the city of rugs on the old caravan trail, there are ruins of a very old Mosque. The bricks just under the cupola have a lot of circles on them. When I asked why, the guide answered "Oh, at the time when the Mosque was built nearly a thousand years ago, the trees had been cut down all around here. In those days we used to build Mosques of wood and these circles, which represent the annual rings, are the only remains of the forests that covered this area, which is now desert."

We all know the figures, the estimate that about one tenth of all plant species on earth are dangerously rare or under severe threat and that this means some 20,000 species may disappear or have disappeared from what was once a very green planet.

When I was a boy, all Swedish schoolchildren had to collect plants and make a little herbarium during the summer vacation. When school started again in the last days of August we showed the results to our teacher and he scrutinized them. They should have had Latin and Swedish names, information on where they had been collected and so on. There were many practical jokes about this, such as borrowing a herbarium from an older brother, so cheating the teachers. But still my generation and some younger ones really learned about flowers in the Linnaean spirit. I think that the present generation of schoolchildren has lost something when they are not forced to learn the flowers of their country.

Collecting of course is no longer the right way to do this; many species were badly diminished by schoolchildren's collections. All the grammar schools that existed in Sweden in the early 1900s had what was called a 'herbarium normale', showing most of the Swedish flowers. I had the honour of being the curator of the collection in my school for two years. The curatorship was passed from generation to generation of schoolboys. That task gave me an inspiration and a lifelong dedication. The Linnaean heritage is still living in Sweden.

Then, some years ago when at last we had started World Wildlife Fund Sweden, I was asked to be the Secretary-General. I will never forget the first meeting with the Projects Committee (or Scientific Committee). The chairman was and is the head of the National Museum of Natural History, Naturhistoriska Riksmuseet, and an expert on fishes. On the committee are Professor Olov Hedberg and Måns Ryberg, the scientific advisers of the Sportsmens Association and the Nature Conservation Society and the head of the Department of Wildlife in the Environment. Now the chairman said at our first meeting of this committee: "Our projects, national and international, are very much centred on the big animals. We must accept that because they appeal to the general public. We must go on working and using the big animals as the ensign of the ship, but we must also sail the ship. And remember, gentlemen, IT ALL DEPENDS ON THE GREEN." Professor Hedberg and I talked about this and the result was our Project Linnaeus— how to save the endangered plant species of Sweden. The project, which was an initiative from WWF Sweden, is now a team effort under the leadership of Dr Ö. Nilsson. The team includes the Botanical Institution and Garden of the University of Uppsala, the National Board for Protection of the Environment, and the National Museum of Natural History. The WWF Sweden gives money to these institutions, organizations and so on to make it possible for their scientists or other people to work with the project. As a principle the WWF try to avoid paying their salaries; they must be paid in some other way. When the project started WWF Sweden also expressed the opinion that as many amateurs and voluntary organizations as possible should be involved in the work.

It was originally estimated that about 300 of the 2000 plants in Sweden were more or less in danger. The first inventory, however, reduced the initial estimate to about 170 species. A questionnaire was sent out to institutions, voluntary organizations and amateurs, especially to places and persons who were supposed to have botanical knowledge. Most species in Sweden are quite well known, especially as to where they occur, how many there are, etc., and so what we needed was often to follow up the present state of knowledge. In many cases the leadership of the parts of the project then went to the actual places that had responded with the original data.

The second phase was to go to the places from where species had been recorded and to investigate their status on the ground. The third phase was to look into the possibilities for the survival of the species in those locations. Then the fourth: if the species was very much in danger, to consider the different conservation actions available. There has been published so far in *Svensk Botanisk Tidskrift* the conservation status for 79 of the so-called endangered species. It is hoped to complete this series by 1979. A book will be published on the subject, a book which appeals to the general public and not just to the specialist. For some species, especially those whose natural environment has been destroyed, it is necessary to bring them into custody in the botanic garden. A centre for that has been arranged in the Botanical Garden of Uppsala. The intention is that each species shall be cultivated in two of the major botanic gardens, two in order to prevent a complete loss of the species if a major accident happens at one garden.

This is what has happened so far with Project Linnaeus. But what about the future? In Sweden and in other countries we have used the law to protect species as a tool for the survival of both plants and animals. Legal protection is especially important in the Scandinavian countries because the laws on rights of access mean that one is allowed to walk wherever one wants (except in cultivated fields and other plantations) and allowed to pick flowers, mushrooms, or gather branches to make a fire. It is very difficult therefore to protect scarce and beautiful plants. Giving plants legal protection has been to translate the delightful Swedish expression "Give them the light of peace" into reality; this policy has been very successful with protecting the popular spring flowers like Cowslip (*Primula veris*), Pasque Flower (*Pulsatilla vulgaris*) and some others. It has also been possible to give at least some protection to the rarer orchids as well as to coastal plants, since people very often walk on the beaches and simply by ignorance pick and destroy important wild plants. There is a tendency amongst scientists to say that it is no use protecting a plant or a flower in this way because what is necessary is to protect the habitat. This of course is true, but there are two snags: first it is very important when one tries to obtain publicity for the plant kingdom to have some pegs on which to hang the conservation message. Scarce and beautiful flowers are a very good peg. A second snag is simply that one can never protect enough habitats.

With the protection of a species as a background it is possible to interest land-owners, communities and so on in giving if not a legal then an interest in the protection of the actual habitat. It is very important that in every country, in every continent all over the world, representative reserves and National Parks be made for all types of habitat. There is a tendency in all discussions among conservationists and scientists to think that it is necessary to have big areas. I feel, however, that it is just as important to have many spots just like measles all over the land. These should not only cover all the different biotopes but also places where you just find one individual species which may be very scarce; these are the sites it is worthwhile protecting with a little nature reserve, a little nature monument or whatever one may call it. From these spots a species can recover and re-invade lost territory. And why is that so important? For me it is because I think it is our duty to give to our children what we have received from our parents. These spots of measles are especially important today when the methods of agriculture and forestry are changing rapidly. Sometimes I feel that the cause of conservation has gained so many supporters because the changes in these industries and occupations have affected the landscape to such a degree that we all feel something has been lost.

Too little is said about the conservation of man-made habitats, for example a hedge or a country road in England, or a hill slope grazed by cows in the Alps, or a meadow under the oaks, elms or lindens in southern Sweden. Saving a man-made habitat needs much more co-operation between people interested in history and tradition, and those interested in nature. Project Linnaeus should be involved in these questions; I should like to see not only conservation of natural habitats, but also the conservation of man-made habitats in co-operation with those people interested in culture and re-creation of such habitats. The more I try to learn about animals, birds, insects, reptiles and fishes the more I understand that all preservation depends on the conservation and re-creation of our biotopes.

In this connection there is one special danger just around the corner. It is the use of all kinds of fertilizers, pesticides and other pollutants. We have in Scandinavia, especially in Norway and Sweden, a specific problem in this respect. Our waters are very much damaged by the precipitation of sulphuric acid borne by the prevailing south-western wind which also brings the rain. Thus we receive sulphuric acid from the big industrial areas in England, France and Germany. All fresh water in lakes, ponds and rivers is becoming more and more acidic. Already the fishes are gone in many lakes and when will the plants in the woods and the meadows follow? Or have they already followed? The need for an international convention in this field is overwhelming and immediate.

As mentioned above the first phase of Project Linnaeus will be finished this year. Now is the time to decide what we should do in the future, the priorities for future action. The first priority is the very rare species whose habitats are endangered. In most cases we already have nature reserves or corresponding protection. It may be necessary to buy land to create new reserves for some of these plants. For some species legal protection may also be needed and this may help if a new locality is found outside the earlier reserves. Very often it is also a question of management. Many of these species are found in more or less man-made habitats where changes in forestry and agriculture have an influence.

The second category are the rapidly declining species. They are very often dependent on a certain traditional way of managing the land. WWF Sweden has just given some money for an investigation of weeds maintained by old agricultural methods and is preparing to support some farmers on Gotland. It is hoped that they will use their land in the traditional manner in order to keep some of the weeds in the Swedish flora. The third category are the endemic species for which every country has its own responsibility. The Swedish endemic species are as a rule not endangered but we must continually monitor them and be prepared to ensure that nothing happens to them.

To carry out a programme like this it is necessary to have money, manpower and knowledge. By the dedicated work of Dr Nilsson and his team I think we now have the knowledge. So it is now necessary to convey a message to the general public in order to get money and manpower; this also needs money. The money may come from the taxpayer or by fund-raising. In most cases it is necessary to make the general public understand the value of a rich and diversified wild flora. We may use film, television, pictures, postcards. Small illustrated articles in the press depicting one flower and telling a story are I think one very good approach: where you find a plant, why it is scarce and the beauty of it, the use, medicinal value and so on.

It may be said that the British still rule the world because of the introduction of their 'lawn imperialism'. All over the world men suffer the burden of keeping a more or less formal garden. In our country many people have summer houses and so we are publishing a book, of which the title will be *Our own little piece of land*. We hope to print 150,000 copies, which in the UK would correspond to about 1,250,000 copies. We will try to tell the story of how to keep the small piece of land around the summer cottage or around the villa in the suburb and so on, as a natural, perhaps man-made habitat. We will try to teach people how to keep it as it is instead of getting a lot of rich soil, creating a lawn with roses and bringing in other non-native elements. We will try to teach people how to keep the old fashioned meadow, the hillside with pine trees and so on. And of course to ensure that no pesticides are used. In order to keep our rich and diversified flora it is necessary to involve people in the conservation and re-creation of suitable habitats. It is also necessary to create a local pride in a certain kind of habitat. We have for instance used the Black Vanilla Orchid, *Nigritella nigra*, in the same way as WWF International has used the tiger as a symbol for the jungle. *N. nigra* is a pretty little orchid which in Sweden only occurs on the limestone soil in Jämtland in the middle of the country; it is a representative of the old lush meadows which now are disappearing because they are no longer economic for producing hay or for grazing cattle. WWF Sweden has supported a scheme to protect and manage one or two of these meadows in all communities of the province. We have found that one of the best ways to get real conservation work done is to create local pride in the local community or the province of the country. If you build that pride it is very often quite easy to also get the money for the work you want to do. Councils and parliaments are much easier to handle when they have the support of public opinion, which is based on the knowledge that the people are owners of something beautiful, rare and endangered.

There is of course no difference between the raising of money for plant conservation and other fund-raising. There is however one big difference between animals and plants in this respect. With animals we can use movement, voice, sculptures or other corresponding things as a fund-raising tool, but with plants we can only use pictures in various different ways. It is the beauty, especially of the flower, that is the most important way of conveying our message. "A thing of beauty is a joy for ever." When one way or other we have got some money to spend, the spending must be based on knowledge; if we do not have that knowledge we must spend money to get some specific, scientific information on those projects we want to support. It is also necessary to try to go step by step. I think our Project Linnaeus is a very good example of this principle. First inventory, then investigation of habitats, then conservation measures including creating reserves, management support and the botanic gardens. Of course every species ought to be saved in its natural habitat. Sometimes however this is impossible or may only last for a very short time before extinction. Nevertheless, very often botanic gardens cannot provide the conditions of the natural habitat and also it is impossible for botanic gardens to cultivate all endangered species. There must be priorities; after decision based on knowledge an adequate number of individuals must be brought into botanic gardens (at least two of them) as a matter of security. It is better to take material for research from this gene bank rather than from nature. Of course the botanic gardens must try to increase the number of individuals of a rare species. Sometimes it may be possible to re-introduce them back into the wild when the conditions in a certain habitat are better. To take care of species in a botanic garden is the 'last resort' action and sometimes a 'fire brigade' action. But the basic principle must be to conserve the species in their natural habitat. The collecting must never lead to endangering the plants in the wild. I understand that there also is a danger that plants reproduced in a collection may differ from what would be expected if evolution took place in the natural habitat.

But the saving of endangered plant species in botanic gardens is only a third issue for the big gardens. The first is to keep stocks for scientific work, the second to educate the public about the importance of plant life and conservation.

My conclusions therefore are "IT ALL DEPENDS ON THE GREEN"; this must be a motto and rule for all conservationists. We have the great responsibility to give to our children what we have received from our parents; because of this it is necessary to save those plants from extinction that we have endangered, very often by our greedy exploitation of nature. We can only save them on a basis of sound specific and scientific knowledge. In order to do that it is necessary to get money. We do not get the money if we do not succeed in creating in the general public an awareness, responsibility and pride of the rich and diverse world of plants. In order to save the endangered species we must get more conservation areas and nature reserves as well as support management of habitats. At present perhaps conservation is too much involved in conservation of virgin areas and too little re-creating or supporting man-made habitats. We should do our utmost to save rich and diversified plant life in the natural habitat. The botanic gardens have their responsibilities. They may save endangered species and provide a gene bank for them. They should supply scientists with living material from scarce and endangered species. They have a very important rôle in creating public awareness and pride in our rich and diversified flora.

For we must take two individuals of each plant with us to the future – and some bumble bees.

The Rôle of the Nature Conservancy Council in the Conservation of Rare and Threatened Plants in Britain

D. A. RATCLIFFE

Nature Conservancy Council, London

The Nature Conservancy Council (NCC) is the British Government's official agency for nature conservation, with an annual budget of about £7,000,000 and a staff of about 650. It has developed a two-pronged strategy in performing its executive functions, by promoting the conservation of the most important wildlife areas and physical features of Britain, principally as National Nature Reserves; and by seeking to influence all other users of land and natural resources through an educational and advisory capacity, so that a nature conservation ethic is developed and applied more widely. To support these functions the NCC conducts and commissions research to enlarge and enhance the strategic base of scientific information and to give direct insight into the ways of manipulating habitats, communities and species towards desired ends.

Species conservation, both plant and animal, is an integral part of all three aspects, and conservation of rare and threatened species is particularly a matter of concern. The NCC has recently conducted a countrywide survey to identify the specific areas of greatest biological importance to nature conservation, and has published an account of 735 of these together with a background description of the natural heritage of wildlife and habitats. This *Nature Conservation Review* pays considerable attention to the conservation of the native flora. Two of the major criteria applied in the evaluation of sites are *rarity* and *fragility* (=intrinsic sensitivity+threat) of ecosystems, and they can be used similarly in evaluating species most in need of conservation. A basic precept in the NCC's strategy is that special effort has to be made to safeguard that which is otherwise most likely to disappear.

While Britain has many nationally rare and threatened plant species, on the international scale the numbers in these categories are very small. The Council of Europe *List of rare, threatened and endemic plants in Europe,* covering only vascular plants, shows that Great Britain has only 19 species which are rare and threatened on the European scale, and only 15 endemic species. Britain and Ireland (in which NCC has no responsibilities) are more important internationally for their hyper-oceanic floras, especially of ferns, bryophytes and lichens, and particularly in western districts. They have probably the richest assemblage in Europe of this interesting group of plants, which includes some endemic species and a much larger number with an extremely restricted and/or disjunct world distribution. There is also a general abundance of some Atlantic plants which have a somewhat local and western distribution on the European mainland,

such as *Endymion non-scriptus* and *Ulex europaeus*. The British flora is also
of considerable interest in showing a unique combination of phytogeographical
elements, e.g. the mixture of southern Atlantic and Arctic-alpine types in the
extreme west, and the special blend of various groups in the Upper Teesdale
assemblage.

The Nature Conservancy made an early start to surveying the vascular flora
of Britain and Ireland by a grant to the Botanical Society of the British Isles to
carry out a species distribution mapping scheme, using the services of the
membership, mainly amateurs working in their spare time, to do the necessary
field work. This led to the publication in 1962 of the *Atlas of the British Flora*
by F. H. Perring and S. M. Walters, followed by a *Critical Supplement to the
Atlas of the British Flora* in 1968. The mapping scheme was based on the now
widely accepted conventional system using dot records for presence in each
square of a grid, in this case the 10×10 km squares of the British National
Grid. The recording system and the data became the foundation of the Biological
Records Centre, which was taken into the Nature Conservancy and expanded
its work to include the recording and distribution mapping of a wide range of
the major taxonomic groupings of plants and animals. Schemes are operating
for mapping bryophytes, lichens, Charophyta, and certain groups of fungi and
algae, mostly under the aegis of the societies concerned.

The dot distribution maps give a reasonably objective way of identifying
different abundance classes, though the initial definition of their limits is
essentially arbitrary. Dr Perring has used occurrence of a species in 15 or fewer
grid squares after 1930 as the criterion of a 'rare' vascular plant, and has thereby
drawn up a list of 321 taxa (species or subspecies) of vascular plants in Great
Britain, representing about 18 per cent of our native or probably native flora.

The mapping scheme also gives the basis for monitoring, by repeated survey
to reassess distribution. Separation by different symbols into earlier and recent
records (before and after 1930) was also used in the first edition of the *Atlas of
the British Flora* to show changes in distribution, these being mostly of decline
which can be matched against major forms of human impact, especially intensification of agriculture. The method thus allows periodic updating of maps, and
a revision of the *Atlas* taking account only of the rare species has already been
made (1976), while the Pteridophyta have been revised as a group. The study
of change in distribution allows one to identify the threatened as well as the rare
species, i.e. those which are declining most rapidly. The combination of the two
criteria is a measure of the nature conservation importance of species, or more
particularly of the urgency for action in safeguarding them. Dr Perring and
Miss L. Farrell have further refined this assessment of threat in the *British Red
Data Book: Vascular Plants* (1977) by combining rate of decline, number of
extant localities, attractiveness and exposure to collecting, percentage of localities
in nature reserves, and accessibility in terms of remoteness and physical site
factors. These categories can be quantified, and together give a composite threat
number. The *Red Data Book* thus gives the necessary strategic information on
which to base a conservation programme.

Since the major threat to species' survival is the various aspects of human impact, the basic problem is one of finding the best methods of controlling or ameliorating human activities and their effects on the environment. Most of these activities are concerned with orthodox economic advancement, so that it is important to promote nature conservation as a valid resource use in its own right. The management of selected areas as reserves for wildlife does not necessarily involve conflict with other land use, and the presence of many species of plants is dependent upon certain traditional practices, e.g. the grazing of many permanent grasslands, or their use for hay-making. The conflict usually arises when there is an intensification of methods in order to maximize crop yields.

While National Nature Reserves (NNRs) have not been chosen specifically to safeguard a single rare species, the presence of an aggregation of rare or local species has been an important factor in the selection of these reserves. Moreover, areas chosen as NNRs for their important examples of major ecological formations and plant communities are often rich in rare species too, so that these will tend to be well represented as a matter of course. If the 420 grade 1 and 315 grade 2 sites identified in the *Nature Conservation Review* can be safeguarded to the standards obtaining on NNRs (162 of them are already established as NNRs in whole or in part), then c. 78 per cent of the rare vascular plant species would be represented on at least one of these key sites. Most of the remaining 22 per cent belong to unstable and artificial habitats which are not readily protected by nature reserve status.

In addition to the NNR series there are many other nature reserves established by local naturalists' trusts, and often chosen specifically for the presence of rare and threatened plants. The reserves of the Royal Society for the Protection of Birds, and the extensive properties of public bodies such as the Forestry Commission, the Ministry of Defence and the two National Trusts also contain a good many rare and local plants which can be conserved by drawing the attention of these custodians to them and advising on management of the habitats concerned. A good deal is being achieved in this way. The NCC has also notified to local planning authorities and land-owners about 4000 Sites of Special Scientific Interest (SSSIs), many of them recognized as such from the presence of rare species. These SSSIs have no statutory safeguards, but any proposed change in land use under the planning laws must be notified to the NCC who are able to comment on the effects of the proposals, and if necessary make objections which may be upheld if the planners judge that an adequate case has been made. Amongst the habitats actually created by man, road and railway verges are important for their flora, which includes a number of rare species. Notification of important sites to the highway authorities and British Rail, along with suitable advice on management, is having a beneficial effect in conserving many populations of the rarer plants in these habitats.

Some of the research conducted by the NCC has been aimed at understanding the management of habitats and species, and autecological studies of selected rarer species have been made, e.g. *Anemone pulsatilla* (= *Pulsatilla vulgaris*) and *Spiranthes spiralis* in chalk grassland. Further work on rare and threatened plants is in hand, as on *Cypripedium calceolus*, and more is intended. The

Biological Flora series of the British Ecological Society is also an important source of autecological information relevant to rare species requirements. From knowledge of the specific ecological requirements of such plants it is possible to manage their habitats to best advantage on nature reserves, and to advise other owners and occupiers of land with rare species similarly.

A large part of the NCC's work is educational, showing the concerned public the importance of the natural heritage of wildlife, and instilling a sense of the need to cherish this and to act in trusteeship for generations to come. We have to infuse ecological awareness and the notion of avoiding depletion of renewable resources, especially by thoughtless actions such as collecting and other activities causing damage in rare plant localities. NCC's booklet *Nature Conservation and Agriculture* points to the conflict between wildlife and modern intensive farming, and shows how some of the most threatened species are the once-common 'weeds' of cornfields, such as *Agrostemma githago*, which is almost extinct in Britain through the cleaning of cereal seed before planting. The former practice of allowing land to lie fallow, which benefited many arable farmland plants, is now a luxury that few farmers can afford. This illustrates the need for special arable weed nature reserves, and attempts are being made to establish these, and to maintain populations of the dwindling species, which include several of the Breckland annuals. *Nature Conservation and Agriculture* shows how action can be taken to reduce the problem of impact of modern agriculture on wildlife, including the provision of detailed advice to farmers, but identifies the inevitability of the conflict. The nation thus needs to face this squarely and to make adequate dispensations, notably through funds to compensate economic shortfalls which occur when farmers forego the maximization of crop yields in the interests of nature conservation. There is also a need for a national land use strategy which would include promoting the conservation of those areas identified as outstandingly important for their wildlife.

Similar problems exist in regard to modern forestry practice, in which many native, broad-leaved woodlands have been converted to conifer stands, and the bulk of the new forests extensively planted on open moorland and heath are of alien conifers. Afforestation of open ground causes gross habitat changes, and is causing loss of rare and local species especially on wet ground.

Legislation has recently been used to give general protection against picking and uprooting of wild plants in the countryside, and should help to reduce depredations on common or local and attractive flowers. In addition, 21 rare and attractive vascular plants have been given special protection against collecting, since some species which are secure against almost every form of incidental threat are still much at risk to deliberate collecting. This is particularly true of certain montane species and especially the rarer ferns. There is still a problem here over prevailing attitudes to collecting, and this remains one of the biggest threats to rare species. It is unfortunately true that the best defence is often total secrecy over localities, when these are not already well known. My own view is that the chief benefit of legislation is in moulding the climate of public opinion about the value of wild plants.

The NCC are expanding their research programme to give more comprehensive knowledge of distribution of the less common species of the British vascular flora, thereby enlarging the strategic base of biological records. A pilot scheme, which is described more fully by S. M. Walters (see pp. 38–41), has been set up through a contract to the Cambridge University Botanic Garden, to make a complete survey of the rare plants of East Anglia. This has involved a search for information on known localities from the literature, manuscript data and herbaria, and through verbal contact with botanists, followed by extensive field checking of the accumulated records. The aim has been to determine whether a species is still present in a given locality, and if so, to measure population size and present trends in numbers and abundance. Localities are recorded on large scale maps, and are supplied with supporting information to NCC Regional Officers to file as background information against planning applications or other potential threats. Each rare species is, whenever possible, propagated in the Botanic Garden, and seed is provided for the Kew Seed Bank, so that there is a stock of living material for various scientific purposes, and for re-introduction to the wild if this is thought appropriate and satisfies existing policy on this difficult topic.

It is hoped that similar projects can be developed with other botanic gardens to extend cover of such work to the whole of Britain. Dr Perring has also been involved with the development of regional data centres, notably in County Museums, which will be local sources of information on species distribution and surveys, and he has already organized many population censuses of rare species through the Biological Records Centre, now within the Institute of Terrestrial Ecology.

The work of preparing county or regional Floras continues, often through the efforts of dedicated amateur botanists, and gives the most freely available source of information on the distribution and status of wild plants. Nowadays, these Floras usually include a distribution mapping scheme, on a larger scale than the national system, and sometimes down to the level of 1 km grid squares. The Botanical Society of the British Isles gives close support to these projects, and has its own system of vice-county recording, with designated members acting as recorders, and more notable records published in their journal, *Watsonia*. Much of our knowledge of plant distribution in Britain depends on the field work of the large body of enthusiastic amateur botanists, and an important function of the voluntary nature conservation organizations is to co-ordinate this talent to give optimal results and back-up to the agencies with executive responsibilities in this field. The NCC has worked towards an extremely close relationship with the voluntary bodies for nature conservation, and places great value on their work and support.

It is hoped that, through these various approaches, the conservation of the British flora can in future be dealt with systematically and comprehensively, so as to ensure that populations of rare and threatened species are maintained, and that the accelerated extinction of species through human agency is minimized.

The Eastern England Rare Plant Project in the University Botanic Garden, Cambridge

S. M. WALTERS

University Botanic Garden, Cambridge, England

INTRODUCTION

Since the end of the Second World War, revolutionary changes in agriculture and the spread of urban development have been accompanied, in Britain as in many other countries, by the growth of public concern about the fate of 'wild nature'. The modern nature conservation movement, though traceable back to the nineteenth century, owes its present shape and success to developments since 1949, the year when the Government Nature Conservancy (now the Nature Conservancy Council) was formally constituted. These thirty years have seen a parallel development of the voluntary conservation movement, particularly the Royal Society for the Protection of Birds and the Society for the Promotion of Nature Conservation, the latter as an umbrella organization for the Naturalists' Trusts. In this period the Botanical Society of the British Isles undertook, with financial support from the Nature Conservancy and the Nuffield Foundation, the very successful survey of the distribution of British vascular plants which culminated in the publication of the *Atlas of the British Flora* (Perring & Walters, 1962, 1976). From this 'Maps Scheme' arose the Biological Records Centre at the Nature Conservancy's Monks Wood Experimental Station (now part of the Institute of Terrestrial Ecology), an official documentation system for all the British flora and fauna (see D. A. Ratcliffe's paper on p. 32). By the mid-sixties we had for Britain an unrivalled, detailed knowledge of the occurrence of every wild vascular plant, and the use of this information in policies for the conservation of nature became a matter of real practical concern for both the official and the voluntary conservation movement.

In 1974, following my appointment as Director of the Cambridge University Botanic Garden, I took part in discussions with F. H. Perring, as head of the Biological Records Centre, and representatives of the Nature Conservancy Council, on the feasibility of a pilot scheme, centred in the Garden and directed at the problem of conservation of nationally rare plants within East Anglia, for which Cambridge is a convenient centre. Both my predecessors as Director of the Garden had shown great interest in the native flora and concern about the threats to the continued survival of rare species, and I had been much influenced by their teaching and example. The Cambridgeshire and Isle of Ely Naturalists' Trust had indeed been born in 1956 in the Garden and J. S. L. Gilmour, then Director, was like myself a founder-member and later President of the Trust.

37

Moreover, the range of interest amongst those supervising research in the University Botany School had increasingly involved taxonomic, ecological and cytogenetic studies on British plants; and the new Research Area, established in the Garden in 1958, provided an excellent and expanding base for such studies. For all these reasons, the Garden seemed to be very suitable for a project of this kind.

THE NATURE CONSERVANCY COUNCIL (NCC) CONTRACT

Originally granted for one year only (1974–5), the contract was renewed for a second year, and then, following a favourable assessment of the project in 1976, was put on to a five-year basis. The project, entirely financed by NCC, aims to do two things. Firstly, the Scientific Officer (G. Crompton) prepares a detailed index of all past and present records of nationally rare vascular plants within the region now defined as 'Eastern England' (see figure 1). This work, which is being done on a county-by-county basis, aims to provide the NCC with an accurate, up-to-date, complete and confidential documentation of the exact localities and present population sizes of every nationally rare plant. Using this information, NCC officers can act to prevent or minimize damage and destruction of remaining wild habitats. They can also incorporate information on rare species into plans for the acquisition of new nature reserves or protected areas. Monitoring of the remaining sites for rare species is an integral part of the project: not only is this desirable because of natural fluctuations in population size, but also because insidious habitat changes may be operating with effects that can only be measured over a series of years.

The second part of the scheme is closely linked to the first, and aims to establish in cultivation, where it seems appropriate, reserve stocks of the rare species. For this a special British Conservation Section has been created in the Garden under the Conservation Propagator (R. G. Mellors from 1974–6, now D. Donald). This Section is based in the private Research Area, but is linked to the British wild species demonstration beds in the Ecological Area, which is a feature developed over the last twenty years in the public part of the Garden (figure 2). The present establishment includes an assistant paid for by the grant from Imperial Chemical Industries Pharmaceutical Division (see below). A general view of the Section is seen in figure 3.

The work of the Conservation Propagator falls into three main parts: documentation, propagation, and maintenance of the stocks. The main living collections are clonally-propagated stocks of perennials of known wild origin. About 100 nationally rare plants out of the 321 species listed in the recently-published British Red Data Book (Perring & Farrell, 1977) are recorded from the Eastern region, and of these we already hold representative stocks of 47 species from sites in Eastern England. Annual and biennial species are treated differently from the cloned perennials, and with them our main concern has been to provide from natural populations (where this is not harmful to the survival of the plant) ripe seed for the Kew Seed Bank. In cases where seed is only sparingly produced, the initial seed sample from the wild is grown under controlled conditions to

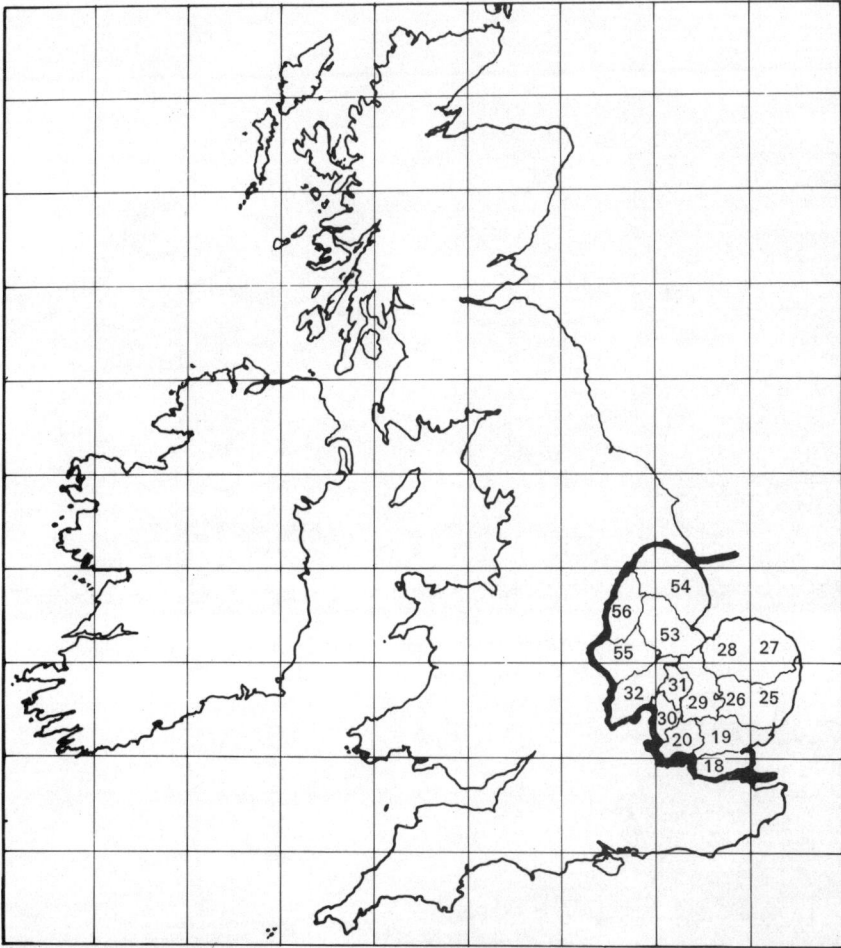

Reproduced by kind permission of the Biological Records Centre,
Monks Wood, Huntingdon.

THE VICE-COUNTY NUMBERS AND CORRESPONDING VICE-COUNTIES

18. South Essex
19. North Essex
20. Hertfordshire
25. East Suffolk
26. West Suffolk
27. East Norfolk
28. West Norfolk
29. Cambridgeshire

30. Bedfordshire
31. Huntingdonshire
32. Northamptonshire
53. South Lincolnshire
54. North Lincolnshire
55. Leicestershire (with Rutland)
56. Nottinghamshire

FIGURE 1.–Area covered by the Eastern England Rare Plant Project

FIGURE 2. The Limestone Mound in the public Ecological Area (All the plants are of known wild British origin). Photo M. G. Walters.

FIGURE 3. The British Conservation Section: general view, with stock beds of rare perennial species in the foreground. Photo M. G. Walters.

prevent accidental cross-pollination, and a much larger seed sample obtained for the Seed Bank. In addition to the main Seed Bank storage we have found it useful to operate our own subsidiary seed bank and to undertake some cultivation of annual and biennial species, mainly for educational display and for research.

Two other groups of native species require special comment. The woody plants, which are relatively few in number, are being treated as a whole, whether common or rare, and the NCC contract allows us to grow documented stocks of any taxa of known wild origin in the Eastern region. This enables us to rehabilitate the very good teaching collection of hardy trees and shrubs already in the Garden, ensuring that the specimen trees of the future will be, where appropriate, of known local origin. In the case of *Salix* (Willow) and *Populus* (Poplar) the taxonomic complexity makes such a new and expertly-verified collection particularly valuable (figure 4). Aquatics and wetland species require special treatment, and we have just erected the first installation to enable us to cultivate at least the majority of the rare species in this group (figure 5).

CONSERVATION PROBLEMS AND THE FLORA OF THE EASTERN REGION

Undoubtedly the most famous group of nationally rare species in the flora of the Cambridge area is the so-called 'Breckland element', well illustrated by the perennial *Silene otites* (figure 6). On a world scale, this and allied species (Sect. *Otites* of the genus) are steppe plants which become increasingly common in Central and Eastern Europe. The presence of a continental, even a 'steppe', element in the flora of East Anglia is closely correlated with a marked continental climate (low rainfall, high summer and low winter temperatures) and a light, sandy soil. Such species are largely (in some cases wholly) confined to the Breckland region of Norfolk, Suffolk and parts of adjacent Cambridgeshire, and have excited the interest of many generations of naturalists. Their cultivation in Cambridge presents no particular problem.

Another group of Breckland rarities are 'winter annuals', adapted to autumn or winter germination and quick flowering and fruiting to avoid the summer drought. Although such species (e.g. *Veronica verna*) are often discussed with the more strictly continental perennials, they differ very significantly in their ecology, and are obviously subject to very drastic population fluctuations which are partly at least caused by changes in land use and agricultural practice. Careful monitoring of rare Breckland annuals has already shown (in the case of the three *Veronica* species *V. praecox, V. triphyllos* and *V. verna*) very differential effects of agricultural and other habitat changes: such studies illustrate well how important is an understanding of the *individual* ecology in the conservation of rare species.

The Fenland species (including the aquatics) present very different problems. Here gross habitat destruction, which began with the drainage of the Fens in the seventeenth century, had already by 1860 (when Babington's *Flora of Cambridgeshire* was published) reduced the possible habitats for most wetland plants to a very small fraction of their former extent. Indeed, had not the two main nature reserves of the Fens at Wicken and Wood Walton come into being

FIGURE 4. The Salicetum in the public area: all the young plants are propagated from
material of known wild origin in Eastern England. Photo M. G. Walters.

FIGURE 5. Installation for growing aquatic and wetland species (in the far left compartment can be seen fruiting plants of *Senecio paludosus*). Photo M. G. Walters.

before the First World War, there would have been no extensive peat-fen area surviving anywhere in East Anglia today. Until 1950 it was generally true that aquatic and wetland plants in the Eastern region were probably not very significantly rarer than they were a century before. Agricultural changes, however, have altered the picture drastically in recent years, and the fate of many aquatic and fenland plants now causes real concern. Against this sombre background it is pleasant to record a 'success story' of a supposedly extinct Fenland plant: *Senecio paludosus* was re-discovered in 1972 and is now surviving, both in the wild, where the population is carefully monitored, and in cultivation in the Garden (Walters, 1974; see also figure 5).

Finally, some comment is called for on the arable weeds, which contain many examples of Eastern England rarities. One of these, *Bromus interruptus*, is apparently totally extinct on a world scale, since it was confined to Britain and has not been since 1972 in its last-known station in Cambridgeshire. Several other weed species are now national rarities; in the case of *Agrostemma githago*, the Corn Cockle, the change from common to rare has come about within a hundred years, and such weeds can no longer survive modern agricultural practices. Botanic gardens would seem to be the obvious places where they can still be grown for display.

In one respect Eastern England is not typical of the British scene as a whole. In such a largely arable region, the proportion of land still under semi-natural (or at least not grossly artificial) communities is at a minimum, and the survival of small populations of formerly more abundant species is generally more

precarious than, say, in the Scottish Highlands. Nevertheless, the procedures
adopted in the project would seem to have general application, although the
urgency of 'rescue operations' may be less in other regions.

VALUE OF THE PROJECT AND FUTURE DEVELOPMENTS

From the point of view of the Nature Conservancy Council, the documentation
of surviving wild populations is of primary importance. Everyone would agree
that the survival of wild plants and animals *in situ* must have the highest

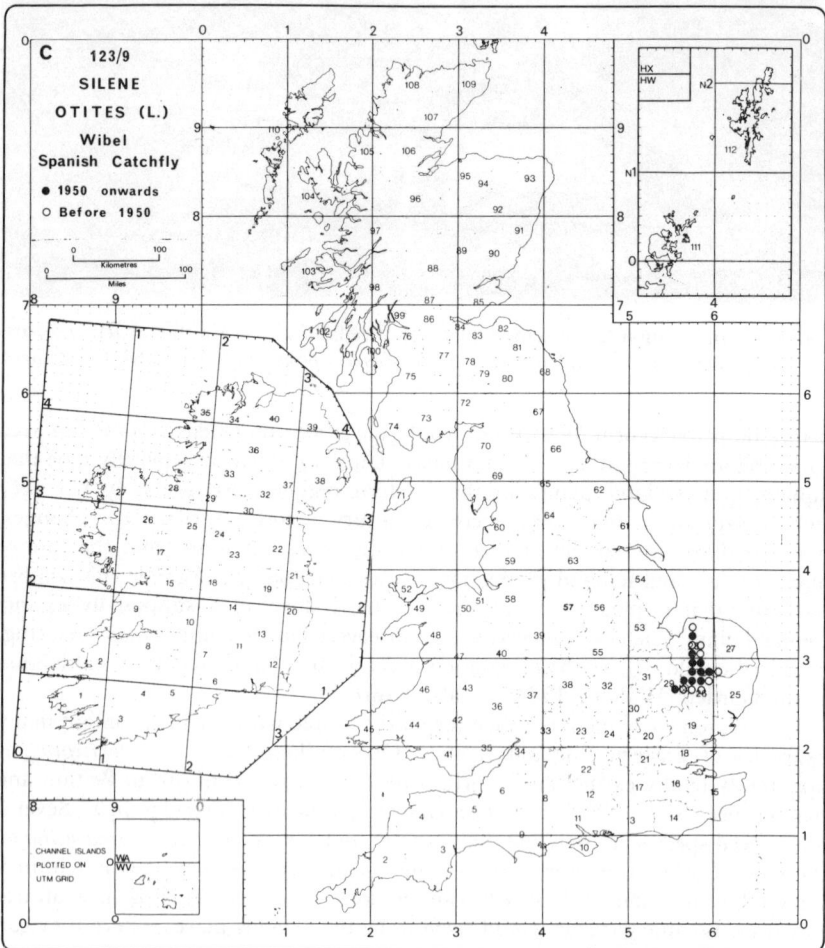

FIGURE 6. Distribution of *Silene otites* in Britain. Produced by the
Biological Records Centre, Institute of Terrestrial Ecology,
from data provided by the Botanical Society of the British
Isles.

priority. Nevertheless, a realistic assessment of the problem must face the fact that some further habitat destruction, whether gross as when a new motorway is built, or more insidiously by pollution effects, is quite inevitable. Holding reserve stocks is some insurance against complete failure to conserve the wild habitats, and at this level alone is worth doing. Since we began the scheme in 1974, however, it would be fair to say that our assessment of the relative importance of holding reserve stocks as distinct from other uses of such stocks has changed and is still changing. From the first, it was apparent that the *educational* value of documented botanic garden stocks of wild plants was very important. There was nothing new in this idea: the present Limestone Mound in the Ecological Area was constructed in 1961–2 using wild-origin British species with such an educational use in mind. What is new is the explicit recognition of *conservation* education. The arrival on the Statute Book of the 1975 Conservation of Wild Creatures and Wild Plants Act in particular provides a very important rôle: it is not practicable to operate such laws unless those who might offend have some chance to see the protected species in cultivation and learn to distinguish them. Beyond this, however, the whole idea of wild plant conservation can be much more effectively explained by means of carefully-designed and labelled 'conservation demonstration beds', and we are turning our attention much more to this use of our rare plant material in the future.

An equally important use which is perhaps less obvious concerns the avail-ability of rare species for legitimate research or other uses. Cloned, cultivated material of wild plants can be used freely for such purposes, thus protecting the remaining populations from further loss. In 1977, the Garden negotiated with the Pharmaceuticals Division of Imperial Chemical Industries (ICI) a contract, originally for two years, now extended to three, to supply samples of taxon-omically verified wild British plants for screening for medicinally active prop-erties in return for an annual grant of £5,000. In the first year of the scheme (1978) the British Conservation Section has fulfilled its contract, and both sides can be said to be very satisfied with the arrangement. Such a function is, of course, a very traditional one for botanic gardens, which were born in the sixteenth century as herb gardens where medicinal plants could be grown and studied by doctors and medical students.

One question which can now be asked concerns the use of the plant collections for ecological research relevant to conservation. We have already mentioned this in connection with the annual Breckland species of *Veronica*. It is certainly my view, and that of the NCC scientists concerned with the contract, that in the future much more use should be made of the material in cultivation to solve specific management problems. Such autecological research is not necessarily sophisti-cated or expensive. A single case will suffice to illustrate this point. When the surviving *Senecio paludosus* stand was found, there was very little ripe seed in the many fruiting heads examined. It seemed possible that self-sterility (common in Compositae) was the cause, since the total population with only three flowering shoots could easily represent a single clone. Subsequent cultivation, and observation of the wild plants over several seasons, have disproved this first hypothesis. Not only are seeds readily obtained, but the germination *in vitro* is usually very high. The species is easy to cultivate, and can be propagated readily both by seed and by vegetative means. Even without any research programme a body of information relevant to the survival of this rare species in nature accumulates in the Conservation Section. For the future, however, more planned programmes, possibly undertaken by full-time research workers, should be attached to the Section. The facilities to develop such research projects in the Garden are particularly good, since the Conservation and Research Areas are adjacent.

Finally, we can ask how far the Nature Conservancy Council is intending to use this pilot scheme as a model for other Regional Centres. This is not for me to answer, but I can say that discussions in other parts of the country have been encouraging, and that similar projects, not necessarily linking both sides of the scheme so closely together as in Cambridge, are already receiving NCC support and encouragement. Like every question in nature conservation, it turns on money and priorities.

REFERENCES

PERRING, F. H. & L. FARRELL (1977). *British Red Data Books: I. Vascular Plants.* Society for the Promotion of Nature Conservation, Lincoln. xxvi+98 pp.

PERRING, F. H. & S. M. WALTERS (eds) (1962, 2nd Ed. 1976). *Atlas of the British Flora.* EP Publishing Ltd., Wakefield, for Botanical Society of the British Isles. xxvi+432 pp.

WALTERS, S. M. (1974). The rediscovery of *Senecio paludosus* L. in Britain. *Watsonia* **10:** 49–54.

A Local Botanic Garden: Its Rôle in Plant Conservation

DAVID BRAMWELL

Jardín Botánico 'Viera y Clavijo', Gran Canaria, Canary Islands

At the first of the Kew Conservation Symposia held in 1975 several speakers, particularly P. H. Raven, pointed out that in areas of the world where conservation problems are most serious, tropical and subtropical regions, there is a general shortage of botanic gardens and active institutions capable of assuming a positive rôle in local plant conservation.

The Canary Islands, with a rich and scientifically valuable endemic flora and many unique vegetation types and plant communities, have suffered from the activities of man probably as much as any other part of the world but, fortunately, very few individual species have been lost completely. Many are, however, reaching a point at which positive action must be taken if we are to ensure their survival. My personal attitude to the conservation of individual species is that unless it can be demonstrated that the species is of absolutely no potential economic, medicinal or any other value for the future, then there is a moral and practical case for trying to conserve it.

Just over 25 years ago a Swedish botanist, E. Sventenius, domiciled in the Canary Islands, realized what was happening to the local flora and proposed the establishment of a botanic garden especially for the endemic Macaronesian flora. He found fortunately, in the local Island Council of Gran Canaria, a number of active and enlightened politicians who gave him their full support. The new garden was founded in 1952 and Sventenius spent the next 20 years dedicated to his project. When he died tragically in a road accident in 1973, he left a garden of some 12 hectares in which was planted about 75 per cent of the Canarian endemic flora.

Since 1974 when I became Director of the garden our efforts have been directed to intensifying its local conservation rôle; this paper attempts to explain this rôle and how we see the future of a local garden in respect to local plant conservation especially in the wild.

First of all, we are trying to do the apparently simple things, and I emphasize *apparently* simple, like getting all the threatened and endangered Canarian endemics into cultivation in the garden in order to bulk up propagating material. We are trying to estimate the size and distribution of natural populations and to establish some degree of protection for them in their natural habitats.

We are also trying to establish a series of small satellite gardens in the different vegetation and climatic zones of the island of Gran Canaria in order to give us a wider range of habitats for difficult plants. We are currently trying to establish a coastal station in the grounds of the Fisheries Technology Institute in the southeast of Gran Canaria for some of the rarest and most problematical plants such as *Convolvulus caput-medusae, Lotus leptophyllus* and *Atractylis preauxiana*, whose natural habitat is rapidly disappearing. Apart from the fundamental task of cultivating Canarian plants, we have managed over the past few years to get together a small but enthusiastic scientific staff of eight graduate biologists, and are now carrying out various conservation-orientated research projects on local plants. These include seed production, dispersal and establishment ecology of coastal populations of endemic species in the Polycarpaeo-Lotetum communities of southeast Gran Canaria, which are threatened by tourist and industrial development, and on the laurel forest communities paying particular attention to the rare endemic understorey taxa such as *Isoplexis chalcantha, I. isabelliana* and *Sideritis discolor*.

Studies on the pollination mechanisms and breeding systems of endangered species are helping us to understand the causes of the natural rarity of plants such as for example *Lotus berthelotii*, which is strongly self-incompatible and, as if to add to its troubles, has a percentage of pollen fertility some 25 per cent below that of other locally endemic *Lotus* species. There are very few clones of *Lotus berthelotii* in cultivation and, though we have large numbers of this plant in the garden, all are reproduced from cuttings of a single clone of known natural provenance. Recently we have been in contact with various European botanic gardens in order to obtain material which may represent distinct clones. We hope to get at least a couple of compatible clones so that we can try and reproduce the species from seed and perhaps prevent fatal genetic erosion in this very valuable ornamental plant.

Lotus berthelotii has the floral morphology of a bird-pollinated flower of the *Clianthus* type. The strong red colour and copious nectar are also adaptations to this method of pollination but the modern avifauna of the islands does not contain any typical pollinating birds. The modern distribution of Sunbirds of the Nectariniidae, however, reaches sufficiently near to the islands on the west coast of Africa, to the edge of the Sahel, to lead us to believe that they may well have formerly occupied the islands; looking at the distribution of *Nectarinia osea* (the Palestine Sunbird) and *Anthreptes platurus* (the Nile Valley Sunbird) on the east side of Africa it is obvious that both the latitude and climate of the Canaries would have been suitable for these birds during and since the late Tertiary period.

The natural rarity of *Lotus berthelotii*, therefore, seems to be due to the loss of its natural pollinators. Other potentially bird-pollinated species in the Canarian flora include *Lotus maculatus, Canarina canariensis, Lavatera phoenicea* and *Isoplexis* spp. These are more common than *Lotus berthelotii*, probably due to the fact that they are all weakly self-compatible and can produce some fruits with viable seeds. We have also been able to demonstrate that *Canarina*, for example, is secondarily accidentally pollinated by insects, especially ants.

In addition we are revising several difficult groups in the islands' flora, trying whenever possible to base the revisions on living material through comparative cultivation, biosystematic studies and observations of variation in wild populations. We now have laboratory facilities for cytology, anatomy and palynology which enables us to cover a wide range of fields. In collaboration with the organic chemistry laboratories of Professor A. González on Tenerife, we can also recommend species for chemical screening for potentially useful products, these recommendations being based upon the findings of our field excursions. Once the plants that are used in folk medicine, etc., are known, we can then provide the research material required for analysis from our living collections.

Apart from cultivation and of course display of endemics and conservation-orientated research, which are the obvious functions of a small local garden, we have over the last few years (perhaps since and indeed due to the earlier Kew symposium) realized that the garden has a fundamental and perhaps its most important rôle to play in conservation education, in order to persuade the local population that conservation of their natural resources in the wild, before it is too late, is a most important factor for the future development of the islands. Our programme in education has been aimed at three different levels: first, schoolchildren; second, their teachers; and third, the general public and their representatives the politicians.

Starting with the general public, our aim is not just that more of them should visit the garden but that they should be aware of and informed about the local flora and the need to conserve it *in situ*. In order to bring local plants to the public attention, we participate in flower-shows, displays, etc., and members of our staff are normally available to give lectures on conservation, the use of local plants in horticulture and so on. We also advise local councils on the use of these plants in public parks and open spaces. Our stands at two local flower-shows this summer were visited by almost 100,000 people. We also publish as many articles as possible on the local flora, especially on its history and economic value, in popular magazines and newspapers in order to reach as wide a public as possible.

Our second target, the school teachers, is an important one, as I am convinced that conservation education must begin with the very young in the classroom; future generations must grow up aware of the issues involved. We have recently organized courses for secondary school teachers on the flora and vegetation of the islands with field excursions. We have been overwhelmed with requests to repeat the courses not only for further groups of teachers but also for tourist guides, local outdoor groups, etc. The purpose of our courses is to enable school staff to make full use of their field excursions and visits to the garden with their pupils, so that we may stress the importance of environmental conservation in general and of the local endemic flora and fauna in particular.

The third objective is to deal directly with schoolchildren especially those aged ten or over. We have an ambitious programme for an exhibition centre and a schools laboratory and we have asked the Island Council for the necessary staff to take charge of such a project. If the response is favourable, we hope to be able to put our programme into effect early in 1979.

A special mention must be made of the future rôle of our type of garden in dealing with politicians who have the power to say yes or no to most of the solutions we can present for local conservation. Rather than becoming obviously involved in politics, which can often have an alienating effect, our rôle at this level is to provide information about the state of our natural communities, the facts on which the politicians can base their decisions. Recently we have been asked to prepare outline projects for two nature reserves on the island of Gran Canaria. The first, for the sand dunes of Maspalomas, has been completed and looks as though it will be accepted by the authorities and the owners. The area is botanically and ornithologically important and we hope it will be fully protected in the future. The second area, the mountain region of Gui-Gui on the west side of the island, is also biologically very interesting as it is the last refuge on the island for *Juniperus cedrus* and for the Osprey (*Pandion haliaetus*). The area is not well known botanically and we will carry out a 12 months survey next year in order to make suitable recommendations for its future.

To summarize paying particular attention to our medium and long term objectives for the future, our aim is to obtain a series of nature reserves on the island of Gran Canaria to cover as many as possible of the local vegetation types and endangered species. We have already made a number of suggestions concerning local areas to be included in a national plan for the protection of natural areas. For small areas with only a single local endemic or for areas too small to be included in a national plan we hope to be able to persuade the Island Council to designate subsidiary local reserves, e.g. for the local habitat of *Euphorbia handiensis*. Thus we hope to have reasonable protective coverage for a majority of threatened plants and areas. It may also be possible to attempt to re-introduce species into their natural habitats within these areas or augment the natural population sizes.

In the botanic garden, if we are to fulfil our rôle in providing material of rare endemics for research and for re-introduction into the wild as part of a reserve management programme, we must be able to maintain a large stock of these species. We would like to devote four or five hectares of land to the cultivation of some of these plants treating them on a semi-agricultural scale with very large plots of each. Starting with as wide a sample as possible of the natural genetic variability of the species, we may at least be able to maintain a high proportion of the gene pool, especially if pollen and seed can be collected from the wild on a regular basis to replenish the garden stock. We would also like to establish facilities to be able to bank both seed and pollen, which would also help prevent genetic erosion in the garden stock of local endemics. A third important aspect of the work is research into the potential value of local endemic species, especially of the legumes as fodder crops; our cytogenetics unit has already started work on *Lotus, Teline* and *Chamaecytisus*.

Undoubtedly this is an ambitious programme but now that we have initiated our conservation drive we must try to keep up the momentum. One of our main problems has been a local shortage of qualified garden staff, especially for example at the curator or deputy curator level. As an exercise in international co-operation between botanic gardens and to help us with this problem, the Zürich Botanical Garden jointly applied with us to IUCN for financial assistance to enable a well-trained and qualified Swiss gardener to work at our garden for a year or two to help us establish propagation units, cultivation techniques, etc., for our endangered plants. Our request for a relatively small sum of $5,000 for a good, solid plant conservation project was basically approved but we were told that unfortunately no money was available. It is perhaps a pity that we have not got the odd cuddly mammal in the garden to pluck at the heartstrings of the IUCN and WWF fund-raisers and givers in order to support this essential development (figure 1).

FIGURE 1. Biological Conservation

It is a great pity, in our case, that we cannot get a little international support from IUCN/WWF because some form of practical international recognition would help our local campaign for more money and staff and, even more importantly, would help convince future generations of politicians that the original decision to found such a garden, taken by their predecessors over 25 years ago, really was an inspired one. With more international support in the future it is quite possible that a garden such as ours could take on some of the responsibility for areas outside Macaronesia. In the whole of the southern Sahara and Sahel region as well as Somalia there are reports of ecological disasters, catastrophic changes in climate and vegetation, destruction of whole vegetation types and so on, but there is not a single centre capable of attempting to save even a few potentially useful plants or very rare local endemics. We have suitable climatic conditions and up to 30 hectares of land available. With international finance for staff we could participate in a plant rescue project for these regions; well adapted local species will be necessary if we are ever able to attempt to reclaim land back from the desert.

The garden could also hold stocks from other islands of the world and grow subtropical and dry-tropical research material for European researchers, but such an international rôle would depend entirely on practical international financial support.

So in summing up the rôle of a local botanic garden in conservation, and using our own garden as a model, it is to play an active part in several areas of the subject:

1. To get into cultivation as many as possible of the local endemics in as large a quantity as practicable.

2. To carry out conservation-orientated research on reproductive biology, establishment, dispersal and also practical uses of local endemics.

3. To play a fundamental rôle in conservation and environmental education at all levels.

4. To provide technical information on which politicians can soundly base their decisions.

5. To play any international rôle designated to the garden and adequately financed.

If we can carry out these tasks successfully on a local scale then we are justifying the existence of such botanic gardens with a definite purpose, and also justifying the expense involved in maintaining a local garden even in poorer areas of the world, where in fact the need for local conservation of local plant resources is most urgent and often least considered.

The Relationship between the National Reserves and the Activities of Botanic Gardens in Plant Genetic Resource Conservation

B. A. MOLSKI

The Botanic Garden of the Polish Academy of Sciences, Warsaw

In 1974 a new botanic garden was created in Warsaw mainly for plant resource conservation. The garden will eventually cover 240 hectares and the first 100 hectares are currently being developed. The garden is the realization of a five year programme sponsored by the Polish Academy of Sciences, entitled 'Plant Genetic Resources Conservation and Utilization', with the aim of establishing an *ex situ* gene bank of all endangered species of the Polish flora. In the programme three other botanic gardens are co-operating with our institution, namely the botanic gardens in Lublin, Wrocław and the arboretum in Kórnik.

ENDEMIC SPECIES

Poland's resources in terms of endemic plant species are poor; these plants are as follows:

> *Erysimum pieninicum* Pawłowski, *Taraxacum pieninicum* Pawłowski, *Saxifraga moschata* Wulf. subsp. *basaltica* Br. Bl., *Viola collina* Bess. subsp. *porphyrea* (Uechtr.) W. Becker, *Cochlearia polonica* Fröhlich, *Galium cracoviense* Ehrendorfer (Szafer & Zarzycki, 1972).

Besides the species mentioned above, there are a few with interesting restricted distributions occurring only in very limited areas outside Poland:

> *Larix polonica* Racib. (Romania), *Betula oycoviensis* Bess. (Czechoslovakia), *Betula obscura* A. Kotula (Czechoslovakia), *Potentilla silesiaca* Uechtr. (Germany), *Carlina onopordifolia* Bess. (USSR).

All these species need to be checked carefully when carrying out further research as their taxonomic status may be in some doubt.

Among the Polish endemic species only two are in danger of extinction: *Taraxacum pieninicum* ($2n = 16$, Małecka, 1958, 1962; Skalińska, 1963) grew on one small rock face in the Pieniny mountains. Some years ago the rock face fell down and since then this species has not been found in the area. Some years ago seeds were collected for cytological and biological studies and plants were grown in cultivation. However, so far there are no plants or seeds available for distribution to other gardens from these collections.

The other endangered species is *Cochlearia polonica* (2n = 36; Bajer, 1950), a member of the Cruciferae. It used to grow in one sandy area over limestone. Several years ago the sand began to be exploited for filling disused coal mines and the species was in danger of extinction. To save it botanists moved the plants to a similar habitat and the species has established there quite well. *Cochlearia polonica* is a very difficult species to grow in cultivation because of its unique ecological requirements. Other endemic species are not endangered and so far are well protected in nature reserves and National Parks.

NON-ENDEMIC RARE AND THREATENED SPECIES

Poland has a long tradition in the protection of endangered species. The first legally-protected plant species was *Taxus baccata*, which was given protection in 1423 when the Polish King, Władysław Jagiełło, ordered this tree to be saved from over-exploitation by the manufacture of crossbows and domestic objects such as spoons, vessels and other utensils (Molski, 1968). In 1919, in the first year of Polish independence, seven plant species were taken under protection: *Taxus baccata, Pinus cembra, Larix polonica, Rhododendron flavum, Clematis alpina, Leontopodium alpinum* and *Schivereckia podolica*. In 1946 a new list of protected species was drawn up by the Ministry of Forestry and Wood Industries; this was expanded in 1957 to 124 fully protected species and 17 partially protected species. This list is still valid and is given as the appendix to this paper.

For fully protected species any collection of plants or seeds from their natural environment is prohibited with severe penalties for law-breakers. Protected plant species are either those of special interest for research or those that are rare and grow in very specific ecological conditions. Some protected species are relicts of glacial migrations, for example *Betula nana, Betula humilis, Salix myrtilloides, Salix lapponum, Linnaea borealis* and *Rubus chamaemorus*; others are interesting biologically, for example insectivorous plants.

The list of 141 protected plant species covers only 6 per cent of the total number of wild plant species found in Poland. However we maintain that about 15 per cent of the species in Poland are in danger, being rare and deserving some form of protection. Some of these species are found in a few, isolated sites; for example *Chamaedaphne calyculata* has only three localities; *Dictamnus albus* grows in two places, *Carlina onopordifolia* has four sites and *Ophrys muscifera* is confined to one site in Pieniny. Only those species which are very distinct and attract attention are under protection. There is little need to protect small inconspicuous plants by law, since they will rarely be collected. The list of partially protected plants also contains species heavily exploited for medicinal or industrial purposes, such as *Ribes nigrum, Atropa bella-donna, Veratrum lobelianum* and *Colchicum autumnale*.

Partially protected plant species may be collected without limitation from cultivated plantations, but even plant material grown for commercial purposes cannot be sold or shipped without a special permit from the District Nature Protection Officer.

Nature reserves are established for many different reasons, but in this paper only nature reserves created for plant conservation are considered. The first nature reserve in Poland was created in 1921, to protect the *Larix polonica* forest in Chełmowa Góra in the region of the Holy Cross Mountains. The next year others were formed in the vicinity of Grabowiec and in Jaksice near Miechów.

Nature reserves for plant conservation may be divided into two main categories: those for protecting a single species and those for protecting a plant community. Table 1 shows the distribution by habitat of the 370 nature reserves (covering 13,948 hectares) that have been created for the protection of indigenous plant communities. Table 2, on nature reserves created mainly for protecting individual species, indicates how the 188 such reserves, covering many hectares, provide protection for 54 individual species. However, it should be remembered that these reserves also protect all the other plant species found within their boundaries.

In addition to nature reserves, seed forests are created for the protection of forest trees, preserving the best populations for seed collection and further distribution. There are 706 such seed forests totalling 5,852 hectares for the protection of the 12 principal forestry species. Specific forest trees are protected in 263 nature reserves covering 6,450 hectares (see Table 3).

There are 528 nature reserves which have been created for the protection of plant species or plant communities with a total area of 20,165 hectares. Among

TABLE 1

Nature reserves formed for the protection of indigenous plant communities as on 1 July 1978

Type of reserves	Number of reserves	Total protected area in the reserves (hectares)
Peatbog	55	3056
Steppes	33	427
Marshy pine forest	4	264
Halophytic vegetation	4	25
Pine forest	9	222
Mixed pine forest	16	448
Spruce forest	9	337
Fir forest	15	720
Mixed broad-leaved coniferous forest	103	4100
Broad-leaved forest	54	1327
Beech forest	46	2293
Oak forest	22	729
Total	370	13,948

Survival or Extinction

TABLE 2

Nature reserves formed for the protection of rare plant species as on 1 July 1978

Species	Number of nature reserves	Total area protected (hectares)
Adonis vernalis	16	212.23
Anemone sylvestris	9	121.93
Anthericum liliago	3	17.25
Aquilegia vulgaris	2	3.77
Carlina acaulis	3	16.28
Carlina onopordifolia	2	5.95
Carex ornithopoda	1	2.04
Cerasus fruticosa	9	37.6
Colchicum autumnale	1	16.67
Convallaria majalis	1	108.54
Cypripedium calceolus	7	130.72
Daphne mezereum	2	9.9
Dictamnus albus	2	36.54
Drosera anglica	9	648.02
Drosera rotundifolia	19	686.58
Drosera intermedia	7	666.99
Epipactis palustris	1	4.06
Gentiana ciliata	1	108.54
Gentiana cruciata	1	21.88
Gentiana pneumonanthe	2	239.57
Gentiana uliginosa	1	4.78
Hedera helix	1	10.73
Herminium monorchis	2	106.00
Iris sibirica	2	43.04
Leucojum vernum	1	4.21
Lilium martagon	3	45.22
Liparis loeselii	1	4.06
Listera ovata	1	39.53
Lonicera periclymenum	1	9.5
Lycopodium clavatum	1	213.2
Lycopodium complanatum	1	87.46
Lycopodium inundatum	1	11.34
Matteuccia struthiopteris	2	39.42
Neottia nidus-avis	1	1.62
Nymphaea alba	3	46.96
Orchis maculata	1	3.00
Osmunda regalis	6	48.1
Phyllitis scolopendrium	1	9.6
Pinus mughus	2	94.33
Primula elatior	1	16.67
Rhododendron flavum	1	0.1
Stipa capillata	5	33.92
Stipa joannis	5	38.42
Trapa natans	4	53.86

TABLE 2 (continued)

Trollius europaeus	2	4.77
Veratrum nigrum	1	108.54
Veratrum lobelianum	1	16.67
Betula humilis	14	872.38
Betula nana	2	420.76
Cladium mariscus	10	258.1
Equisetum maximum	3	3.22
Erica tetralix	5	130.5
Hacquetia epipactis	3	20.92
Linnaea borealis	1	0.17

TABLE 3

Forest nature reserves formed for the protection of tree species and for seed collection

Species	Nature reserves		Seed forests	
	Number of Reserves	Total protected area (hectares)	Number of Seed forests	Total protected area
Abies alba	15	720	64	681
Juniperus communis	1	3	—	—
Larix decidua var. polonica	13	538.01	—	—
Larix decidua	3	13.21	40	199
Picea abies	9	337	123	1372
Pinus sylvestris	25	670	255	1922
Pinus strobus	—	—	2	4
Pinus nigra	—	—	2	6
Pseudotsuga taxifolia	—	—	11	33
Taxus baccata	20	442.62	—	—
Acer pseudoplatanus	—	44	—	—
Alnus glutinosa	27	208	39	181
Alnus incana	—	—	—	—
Betula verrucosa	—	—	2	6
Fagus sylvatica	46	2293	79	722
Fraxinus excelsior	66	710	6	44
Quercus robur	22	729	83	682
Sorbus torminalis	8	223.16	—	—
Tilia cordata	8	119.53	—	—
Total	263	6450.53	706	5852

these are 4 reserves for halophytes (25 hectares), 55 reserves for peatbogs (3,056 hectares), 188 reserves for the protection of individual species and 298 forest reserves (11,793 hectares).

Nature reserves are under the strong influence and pressure from man and his society. In Poland only 8.5 per cent of the area is covered by large natural vegetation complexes, some of them of primary origin (Faliński, 1975), and 33 per cent of the country is without natural vegetation. Peatbogs, xerothermic grasslands and halophytic communities are mostly exposed to degradation from man's activities. Even National Parks are subject to human influence. For example, during the last 150 years almost 40 plant species have disappeared from the Ojców National Park near Kraków, whereas 37 additional species have come in. The Ojców National Park is abundant in plants; more than 1,000 vascular plant species have been recorded in this area (Michalik, 1972). Among species which were once common, but now are completely extinct are *Taxus baccata, Alnus incana, Scopolia carniolica, Prenanthes purpurea* and *Veratrum lobelianum.* Many other species have been reduced in number.

Protected plant species are reduced not only by pressure from man but sometimes because of over-protection of the area in which they are found. Species in steppe communities and in mountain meadows can become displaced by forest shrubs and trees as part of the natural succession. When sheep were removed from the Tatra National Park, many pleasant mountain meadows gradually changed into forest communities reducing the number of many plants, such as *Crocus* spp. Many plant species therefore require selective management, for example by cutting of grass to suppress the growth of shrubs and trees; in fact the protection of many plant species is impossible without man's active intervention in nature.

In Poland there are 614 nature reserves of all types totalling 58,000 hectares. The aim is to increase this figure by 1990 to 950 nature reserves, covering 974 sq. km, that is 0.31 per cent of Poland. In addition to nature reserves there are nature monuments, of which there are currently 7,218, covering 4,954 trees, 1,353 tree groups and 125 tree lanes. By 1990 there should be about 13,000 nature monuments.

NATIONAL PARKS

The best form of plant protection *in situ* is guaranteed by National Parks. The smallest National Park in Poland covers 1,440 hectares (The Ojców National Park) and the largest one 22,184 hectares (The Kampinos National Park). Such areas make it possible to protect whole communities adequately. There are 13 National Parks in Poland (see Table 4) and four more are planned (Table 5). At present National Parks cover 0.53 per cent of the total area of the country, but if their buffer zones are also included, the area covered is about 2,000 sq. km. The national parks of Słowiński, Białowieza and Babia Góra are classified as international biosphere reserves.

TABLE 4

National Parks in Poland

Name	Date established	Area (hectares)	Characteristics
Białowieza	1921	5,072	Lowland forests of primeval character
Holy Cross Mts	1922	6,054	Fir and beech forests on Palaeozoic deposits
Wielkopolski	1932	4,709	Mixed pine-oak forests on moraine hills
Pieniny	1932	2,232	Varied geologically with a unique relief; mountain vegetation.
Babia Góra	1934	1,642	Mountain vegetation on sandstone blocks
Kampinos	1936	22,184	Pine stands on sand dunes
Tatra	1937	21,556	Alpine mountains
Ojców	1954	1,440	Vegetation on Jurassic limestone
Karkonosze	1959	5,508	Mountain vegetation on Palaeozoic rocks
Wolin	1960	4,691	Cliff shores and forests on the Baltic coast
Słowiński	1966	18,068	Vegetation on loose dunes
Roztocze		6,400	Lowland forests
Gorce	1978	6,720	Montaine vegetation

TABLE 5

National Parks Planned

Name	Area (hectares)	Characteristics
Bieszczady	5,900	Montane vegetation
Angustów	20,000	Forest and lakes on moraine
Jaćwieski	16,000	Forest and lakes on moraine
Biebrza	20,000	Peatbogs

There are nine botanic gardens and two arboreta in Poland. The new botanic garden in Warsaw and the arboretum in Kórnik belong to the Polish Academy of Sciences. One botanic garden belongs to a city and the rest are University gardens. Each has a demonstration plot of rare and protected plant species, but there are no special plantations of rare or endangered plant species grown for genetic resource conservation.

In 1975 the Polish Academy of Sciences in Warsaw established a special research programme and allocated 100 million złoty for studies in the biology of rare and endangered plant species. Four botanic gardens (Lublin, Wrocław, Warsaw and Kórnik) have joined in this programme. The genera *Trapa*, *Orchis*, *Cytisus* and *Secale* (wild species) are being studied.

Trapa natans is a protected species which during the last 150 years has been fast disappearing due to river regulation, general land improvement and water pollution. It is estimated that about 75 per cent of all *Trapa* sites have been lost during the last hundred years. Research has shown that it is possible to grow *Trapa* as a cultivated plant in old river beds and studies are now being conducted with the intention of introducing it to ponds connected to artificial lakes. During the next two years, the details will be worked out for the cultivation and fruit storage of *Trapa natans* in botanic gardens and artificial ponds. It is worth mentioning that *Trapa* fruits have been much used as food in the past and that they may be useful in the future for either food or fodder.

The problems associated with the cultivation of *Orchis* in artificial environments and its propagation by tissue culture are well known and recognized in Europe. The botanic garden in Wrocław has been working on this problem for three years.

Cytisus is another genus of scientific interest, due to its xeromorphic character. The botanic garden in Lublin has worked on the group for three years. Field studies are more or less completed and collections are established in the botanic garden. The reproductive biology is now being studied. We hope that in two years time we shall have some information on how it may best be grown in cultivation.

Another group which is of great economic importance are the wild species of *Secale*. During the last few years the botanic garden in Warsaw has collected plant material and conducted intensive research into population biology through isoenzymatic studies and on cultivation in the botanic garden. The assessment of their populations should bring results which will be valuable in the taxonomy of the genus.

It should be emphasized that studies on *Trapa*, *Orchis* and *Cytisus* are carried out with close attention to field autecological observations and experimental work in the gardens. Nevertheless we are aware of many shortcomings in our programme and hope to improve it in the near future.

A list of endangered species is being prepared in co-operation with the Institute of Botany of the Polish Academy of Sciences. We hope to complete this by the end of 1978.

From 1979 we shall undertake a wide research project to establish in botanic gardens such endemic species as *Cochlearia polonica*, *Carlina onopordifolia*,

Galium cracoviense and *Erysimum pieninicum*. A search for *Taraxacum pieninicum* also must be started in order to provide material for further studies.

APPENDIX

The following species are fully or partially protected in Poland:

Trees and shrubs

Pinus cembra L., *Pinus mughus* Scop., *Taxus baccata* L., *Betula oycoviensis* Bess., *Cerasus fruticosa* (Pall.) Woronow, *Chamaedaphne calyculata* (L.) Mnch., *Daphne cneorum* L., *Daphne mezereum* L., *Hedera helix* L., *Lonicera periclymenum* L., *Rhododendron flavum* L., *Sorbus intermedia* Pers., *Sorbus torminalis* (L.) Cr., *Staphylea pinnata* L.

Herbs

Aconitum callibotryon Rchb., *Aconitum gracile* Rchb., *Aconitum moldavicum* Hacq., *Aconitum paniculatum* Lam., *Aconitum variegatum* L., *Adonis vernalis* L., *Anacamptis pyramidalis* (L.) Rich., *Anemone narcissiflora* L., *Anemone sylvestris* L., *Anthericum liliago* L., *Aquilegia vulgaris* L., *Arctostaphylos uva-ursi* L., *Arnica montana* L., *Asperula odorata* L., *Atropa bella-donna* L., *Carex arenaria* L., *Carlina acaulis* L., *Carlina onopordifolia* Bess., *Centaurium umbellatum* Gilib., *Cetraria islandica*, *Cephalanthera alba* (Cr.) Simk., *Cephalanthera longifolia* Huds., *Cephalanthera rubra* (L.) Rich., *Chamorchis alpina* (L.) Rich., *Colchicum autumnale* L., *Convallaria majalis* L., *Corallorhiza trifida* Chatelain, *Crocus scepusiensis* Borb., *Cypripedium calceolus* L., *Dictamnus albus* L., *Drosera anglica* Huds., *Drosera intermedia* Hayne, *Drosera rotundifolia* L., *Epipactis atropurpurea* Raf., *Epipactis latifolia* (L.) All., *Epipactis microphylla* Sw., *Epipactis palustris* (L.) Cr., *Epipactis sessilifolia* Peterm., *Epipogium aphyllum* Sw., *Eryngium maritimum* L., *Fritillaria meleagris* L., *Galanthus nivalis* L., *Gentiana amarella* L., *Gentiana austriaca* A. & J. Kern., *Gentiana baltica* Murb., *Gentiana campestris* L., *Gentiana ciliata* L., *Gentiana clusii* Perr. & Song., *Gentiana cruciata* L., *Gentiana frigida* Haenke, *Gentiana nivalis* L., *Gentiana orbicularis* Schur, *Gentiana pneumonanthe* L., *Gentiana praecox* A. & J. Kern., *Gentiana punctata* L., *Gentiana tenella* Rottb., *Gentiana uliginosa* Willd., *Gentiana verna* L., *Gentiana wettsteinii* Murb., *Gymnadenia conopsea* (L.) R. Br., *Gymnadenia odoratissima* (L.) Rich., *Hierochloë odorata* (L.) P.B., *Herminium monorchis* (L.) R. Br., *Iris aphylla* L., *Iris graminea* L., *Iris sibirica* L., *Leucojum vernum* L., *Leucorchis albida* (L.) E. Mey., *Leontopodium alpinum* Cass., *Lilium martagon* L., *Listera cordata* (L.) R. Br., *Listera ovata* (L.) R. Br., *Liparis loeselii* (L.) Rich., *Lobaria pulmonaria*, *Lycopodium alpinum* L.,

Lycopodium annotinum L., *Lycopodium clavatum* L., *Lycopodium complanatum* L., *Lycopodium inundatum* L., *Lycopodium selago* L., *Lycopodium tristachyum* Pursh, *Malaxis paludosa* (L.) Sw., *Microstylis monophyllos* (L.) Lindl., *Matteuccia struthiopteris* (L.) Tod., *Neottianthe cucullata* (L.) Schlechter, *Neottia nidus-avis* (L.) Rich., *Nymphaea alba* L., *Orchis coriophora* L., *Orchis maculata* L., *Orchis mascula* L., *Orchis militaris* L., *Orchis morio* L., *Orchis pallens* L., *Orchis palustris* Jacq., *Orchis purpurea* Huds., *Orchis russowii* Klinge, *Orchis ruthei* M. Schulze, *Orchis sambucina* L., *Orchis traunsteineri* Saut., *Orchis tridentata* Scop., *Orchis ustulata* L., *Ophrys muscifera* Huds., *Osmunda regalis* L., *Phyllitis scolopendrium* (L.) Newm., *Platanthera bifolia* (L.) Rich., *Platanthera chlorantha* Rchb., *Polypodium vulgare* L., *Primula elatior* (L.) Hill, *Primula officinalis* (L.) Hill, *Pulsatilla alpina* (L.) Del., *Pulsatilla patens* (L.) Mill., *Pulsatilla pratensis* (L.) Mill., *Pulsatilla slavica* Reuss, *Pulsatilla teklae* Zam., *Pulsatilla vernalis* (L.) Mill., *Pulsatilla vulgaris* Mill., *Spiranthes spiralis* (L.) Chevall., *Stipa capillata* L., *Stipa joannis* Cel., *Traunsteinera globosa* (L.) Rchb., *Trapa natans* L., *Trollius europaeus* L., *Usnea* spp., *Veratrum album* L., *Veratrum lobelianum* Bernh., *Veratrum nigrum* L.

REFERENCES

BAJER, A. (1950). Cytological studies on *Cochlearia polonica*. *Acta Soc. Bot. Pol.* **20(2):** 635-646.

FALIŃSKI, J. B. (1975). Anthropogenic changes of the vegetation of Poland. *Phytocoenosis* **4(2).**

MAŁECKA, J. (1958). Chromosome numbers of some *Taraxacum*-species in Poland. *Acta Biol. Cracov., Bot.* **1:** 55-56.

——(1962). Cytological studies in the genus *Taraxacum*. *Acta Biol. Cracov., Bot.* **5:** 117-136.

MICHALIK, S. (1972). Synantropizacja szaty roślinnej ojcowskiego parku narodowego. *Phytocoenosis* **1(4).**

MOLSKI, B. (1968). Gatunki drewna uzywane w średniowiecznym Szczecinie do wyrobu przedmiotów codziennego uzytku. *Archeologia Polski* **13(2):** 491-502.

SKALIŃSKA, M. (1963). Cytological studies in the flora of the Tatra Mts: A synthetic review. *Acta Biol. Cracov., Bot.* **6:** 203-233.

SZAFER, W., S. KULCZYŃSKI & B. PAWŁOWSKI (1953). *Rośliny Polskie*. PWN, Warsaw.

SZAFER, W., & K. ZARZYCKI (1972). *Szata roślinna Polski*. PWN, Warsaw. 2 vols.

The Rôle of Tropical Botanic Gardens in the Conservation of Threatened Valuable Plant Genetic Resources in South East Asia

E. SOEPADMO*

Department of Botany, University of Malaya, Kuala Lumpur, Malaysia

SUMMARY

The tropical rain forest is generally considered as one of the most complex and species-rich terrestrial ecosystems in the world. In S.E. Asia it constitutes the second largest block of such forest in the world and up to 1969 covered a total area of approximately 200 million hectares. Within this forest there are not less than 25,000 species of flowering plants, many of which are endemic and of great scientific interest, while others possess economic potential in terms of the diversification and improvement of local crops. However, due to the ever increasing population, rising standard of living, and demand for raw materials by the developed and industrialized countries, large tracts of primary forest are being logged or clear felled and transformed into agricultural lands. In West Malaysia alone about 200,000 hectares are being logged and cleared annually. This activity, though it has a very sound economic basis, has threatened the existence and survival of many species of plants not found anywhere else in the world.

The present paper attempts to highlight the current problems of conservation faced by developing countries in S.E. Asia and outlines the various steps already being taken to prevent the complete erosion and extinction of many valuable plants. In particular, the rôles of botanic gardens and arboreta as suitable places to conserve living specimens of those rare and threatened plant species will be discussed in detail.

INTRODUCTION

The forests. Tropical rain forest in S.E. Asia has been considered by many botanists and ecologists as one of the most complex and species-rich terrestrial ecosystems in the world (van Steenis, 1951; Richards, 1952; Ashton, 1964; Wyatt-Smith, 1966; Poore, 1968; Jacobs, 1974; Whitmore, 1975; Prance, 1977). It constitutes the second largest block of rain forest in the world and up to 1969 covered a total area of c. 200,000,000 hectares (Pringle, 1969; Whitmore, 1975).

* Read by Dr G. Smith

According to a conservative estimate made by van Steenis (1971), this forest contains not less than 25,000 species of flowering plants. In Peninsular Malaysia, where the land area covers c. 13.2 million hectares, there are about 8,000 species of angiosperms which have been described. Studies carried out by Ashton (1964, 1976, 1977), Poore (1968), Ho (1971), Whitmore (1971, 1973a), Soepadmo & Kira (1977) and many others show that in such forests there are between 400 and 1,000 individual trees with a diameter of 10 cm or more per hectare. Depending on topography and soil type, these trees may belong to 100–200 different species. These studies also indicate that in general each species (except for a few herbaceous plants) is represented by not more than 5 individuals per hectare on average. This means that though the forest is extremely rich in species, the density of individuals of each component species per unit area is extremely low. As a consequence and coupled with the fact that most species are either facultative or obligate out-breeders, a large tract of undisturbed forest is required to maintain a self-perpetuating breeding population of a given species.

The complexity of tropical rain forest is attributable to the presence of the different life-forms exhibited by its component plants. Many plants, e.g. members of the Dipterocarpaceae and Leguminosae, are gigantic trees reaching up to 55 m in height and 2.5–3.0 m in diameter, while others are trees of medium size (e.g. members of the Lauraceae, Meliaceae, Myristicaceae, Myrtaceae, Sapindaceae, Sapotaceae, etc.). Also included are shrubs and small trees (Euphorbiaceae, Palmae, Tiliaceae, Ulmaceae, Violaceae), woody climbers (Connaraceae, Leguminosae, Menispermaceae, Rubiaceae, etc.), herbs (Araceae, Commelinaceae, Melastomataceae, Musaceae, Zingiberaceae, etc.), epiphytes (Orchidaceae, Gesneriaceae, Pandanaceae, Urticaceae), parasites (Balanophoraceae, Loranthaceae, Orobanchaceae, Rafflesiaceae) and saprophytes (Burmanniaceae, Orchidaceae, Petrosaviaceae).

Land and forest resources utilization.   As has been indicated by Wong (1971), Jamil (1973), Cockburn & Hepburn (1973), Lee (1973), Salleh (1973), Pusparajah & Chan (1973), Darus (1978), Reyes (1978), Nalampoon (1978), Daryadi (1978) and many others, the rate of exploitation of natural forests for timber production and conversion of forested land to agriculture and plantations in S.E. Asian countries is indeed very rapid. In Peninsular Malaysia, for example, whereas in 1966 c. 9.1 million hectares of the total 13.2 million hectares of land area were still under primary forests, by the end of 1977 only about 7.2 million hectares remained in such condition. This represents a loss of nearly 2 million hectares of primary rain forest within a 10 year period, or c. 200,000 hectares per annum. Of the remaining 7.2 million hectares of forests, about 2.1 million hectares will be logged and transformed into agricultural land in the near future. Of the 5.1 million hectares of forested land, 3.2 million hectares will be logged and converted into Permanent Forest Estate. This means that by the turn of this century there will be only approximately 1.9 million hectares of primary rain forests left in Peninsular Malaysia, or only about 14 per cent of the total land area. Since forest exploitation and agricultural development in Peninsular Malaysia could be considered as better planned and undertaken than elsewhere in S.E. Asia, the rate of loss of primary rain forests in the neighbouring countries (Indonesia, Philippines and Thailand) may be even more alarming.

This picture clearly indicates to us that primary rain forests in S.E. Asia, in which a large number of flowering plant species are growing, are indeed disappearing very fast. Along with the forest loss many plant species will go for ever, since transformation of forested lands to agricultural plantations involves clear-cutting of all standing vegetation, and under forestry practice not more than 10 per cent of the original number of tree species will be encouraged to regenerate or will be replanted.

VALUABLE PLANT GENETIC RESOURCES IN S.E. ASIA

Many botanists, ecologists and geneticists (e.g. Vavilov, 1951; Frankel & Bennett, 1970; Frankel & Hawkes, 1975; Jong *et al.*, 1973; Sastrapradja, 1975; Burley & Styles, 1976) have pointed out that S.E. Asia is one of the world centres of origin and diversity for many important crop plants (fruit trees, ornamental plants, sugar cane, timbers, yams, etc.). However, due to historical, political and socio-economic circumstances, only a few of the available plant genetic resources have been utilized; the rest are still growing wild in the jungles where their potential is waiting to be tapped properly and rationally. Due to limited space, only those which are considered as important crop plants will be discussed.

Timber trees. Two thousand five hundred of the 8,000 species of flowering plants found in Peninsular Malaysia are trees. Of these, 677 species belonging to 168 genera are able to reach timber size, i.e. at least 40 cm in diameter, and are classified as timber species (Kochummen, 1973; Lee & Chu, 1974). However, under the so-called Malaysian Timber Grading Rules of 1968, only 402 species distributed in 71 genera are considered as commercial timber trees and classified as heavy hardwood, light hardwood or softwood. The remainder are regarded as lesser known timber trees and are very little used. Of the 402 species of commercial timber trees about 157 are dipterocarps belonging to the genera *Anisoptera, Balanocarpus, Cotylelobium, Dipterocarpus, Hopea, Shorea, Parashorea* and *Vatica*. The rest are species of the Anacardiaceae, Annonaceae, Apocynaceae, Bombacaceae, Burseraceae, Celastraceae, Lauraceae, Leguminosae, Myristicaceae, Sapotaceae and a few others. Production of timber during the period from 1966 to 1976 amounts to 15.0 million tons, deriving revenues and foreign currency earnings of not less than MAL$ 300 million per annum. This represents the third largest source of income in the country. In addition, the wood-based industries provided 150,000–200,000 jobs to Malaysians. Thus the contribution of the forestry sector to the national economy is indeed very significant. Unfortunately however, to extract about 15 per cent of the standing trees, c. 50–60 per cent of the basal area of the forest is destroyed (Burgess, 1971). The remaining 25–35 per cent of the basal area of the forest will eventually be treated silviculturally, killing the so-called undesirable (from a forestry point of view) plant species. Since under the current regeneration and replanting practices adopted in the region only about ten per cent of the original number of species of trees will be encouraged to grow, the rate of erosion of timber-tree genetic resources is very high. It has been estimated that at the current rate of exploitation, by the end of this century nearly all lowland and hill dipterocarp forests will have been logged and either regenerated with a lower species content or transformed into agricultural lands. Along with this about 60–70 per cent of the original flora will be wiped out, while the remainder will be found only in small and isolated pockets of virgin jungle reserves. Only those species which can adapt to the new man-made environment will be able to survive outside these reserves.

Fruit trees. Recent surveys by Meijer (1969), Jong *et al.* (1973), Low (1975), Sastrapradja (1975), Valmayor & Espino (1975) indicate that currently there are c. 124 species of fruit plants being cultivated in S.E. Asia. Of these not more than 25 or 30 per cent are native species and the others originated from elsewhere, particularly from tropical Central and South America. In addition to these and as has been shown by Kostermans (1958, 1966), Jarrett (1959, 1960), Soegeng-Reksodihardjo (1962), Soejarto (1965), Meijer (1969), Ho (1971), Whitmore (1971), Jong *et al.* (1973), Hou (1978) and Soepadmo (1977), there are more than 100 species of native fruit trees still growing wild in the forests. Several of these have been known locally to produce edible fruits while others are considered as having great economic potential for breeding and selection purposes. Each of these fruit tree species occur in the forest at a very low density, i.e. five individuals per hectare on the average. The fruits are gathered and consumed by local inhabitants at least once a year. Due to the lack of detailed knowledge of their ecological and horticultural requirements and basic information on their reproductive biology and breeding systems, these wild fruit tree species until now have either been neglected or are very little used. With a few exceptions (e.g. species of *Artocarpus* and *Baccaurea*) most of these species cannot survive well in disturbed habitats such as logged-over or regenerated forests.

Ornamental plants. Though currently there are more than 500 species and cultivars of ornamental plants being cultivated in tropical gardens of S.E. Asia (Holttum, 1962; Chin, 1977; Ismail Saidin, 1977), not more than 50 (or 10 per cent) have originated locally. This is despite the fact that in the rain forests there are hundreds of plant species which have great potential for being introduced and planted as ornamentals. Among these are medium or small-sized trees having either a graceful life-form or producing attractive flowers or colourful young flushes (e.g. species of *Calophyllum, Casuarina, Cerbera, Cinnamomum, Gardenia, Hibiscus, Mesua, Millettia, Pentaspadon, Polyalthia, Sterculia, Symingtonia, Tabernaemontana, Xanthophyllum*, etc.), attractive shrubs and small trees (e.g. palms; *Anisophyllea, Ardisia, Gardenia, Ixora, Pavetta, Rhododendron*, etc.), and herbaceous plants with attractive flowers or foliage (e.g. *Aglaonema, Alpinia, Argostemma, Costus, Curculigo, Cyperus, Cyrtandra, Didymocarpus, Medinilla, Orchidantha, Phyllagathis, Tacca*, etc). In addition there are many ferns and fern-allies which could be introduced as ornamental plants. One of the reasons for the lack of interest by local people in growing native plants with a great potential as ornamentals is that most of these plants are not readily available for cultivation. They grow wild in different types of forest and horticulturally are very little known. In consequence, it is generally more profitable for local nurseries to raise and propagate the already well known and popular plants such as *Acalypha, Codiaeum, Begonia, Celosia, Chrysanthemum, Hydrangea, Petrea, Rosa*, etc., rather than spending time and investing money to collect, cultivate, propagate and introduce unknown native plants to local gardens. In the meantime the population sizes of nearly all wild species of ornamental plants are shrinking very rapidly due to the loss of their natural habitats. Even in the case of orchids, of which in S.E. Asia there are more than 1500 species, not more than 50 are widely cultivated, propagated and hybridized in the region. The rest are being wiped out along with their natural habitats or owing to unscrupulous over-collecting.

Medicinal plants. Unlike the situation in well developed countries where the majority of the population can enjoy free modern medical service or are able to buy laboratory-prepared medicines, in the under-developed and developing countries of S.E. Asia, the majority of the inhabitants still use or believe in curing various illnesses with traditional medicines prepared in their own home and utilizing local plant materials. Dongen (1913), Boorsma (1926), Heyne (1927), Burkill (1935), Steenis-Kruseman (1953), Soepardi (1964) have listed not less than 1000 species of plants which have been used for generations by local inhabitants in preparing various types of traditional medicines. Many of these plants are forest inhabitants and may have completely disappeared or become very rare before their medicinal properties can be examined scientifically. The survival of those which are still in existence today is seriously threatened through the loss of their natural habitats. Examples of plants which may have a great potential for medicine are many members of the Apocynaceae, Araceae, Balanophoraceae, Dioscoreaceae, Lauraceae, Leguminosae, Rutaceae and Zingiberaceae.

Vegetables. As in the case of ornamental plants, out of 90 or so species of vegetables commercially grown in S.E. Asia only about 15 have originated locally (Herklots, 1972). This is despite the fact that according to Ochse *et al.* (1931), Burkill (1935), Sastrapradja & Kartawinata (1975) there are at least 300 species of native plants which have been gathered or cultivated and consumed by local people as vegetables. Of these about 80 species or 26 per cent are wild plants growing in the forests, 95 or 30 per cent grow as weeds, 20 or 6 per cent wild but are occasionally cultivated, and 120 or 38 per cent are regularly cultivated. The fate of these plants in terms of their survival and genetic erosion is no different from that of timber trees, ornamental and medicinal plants.

Tubers and carbohydrates. Accounts by Burkill (1951, 1960), Furtado (1940), Coursey (1972), Purseglove (1972), Whitmore (1973b), Martin (1975), Dransfield (1977b) and Kiew (1977) indicate that in S.E. Asia there are hundreds of species of the Araceae (*Alocasia, Amorphophallus, Colocasia, Xanthosoma*), Dioscoreaceae (*Dioscorea*), and Palmae (*Arenga, Caryota, Corypha, Eugeissona, Metroxylon* and *Oncosperma*) which produce edible underground tubers or young shoots or stems containing a high concentration of starch and sugars. Of these only a few species of *Colocasia, Dioscorea, Metroxylon* and *Xanthosoma* are currently widely and commercially cultivated. The remainder are wild plants growing in the forests at a low density and are either completely neglected or disappearing at a very fast rate along with their natural habitats.

CONSERVATION—THE NEED AND THE PROBLEMS

The need. From the foregoing it is obvious that the need for conserving many valuable plant genetic resources in S.E. Asia is indeed very great. Due to various reasons these plants are disappearing very rapidly and are becoming very rare or approaching extinction before their economic potential can be confirmed scientifically. It is expected by many informed observers and organizations that by the end of this century more than 80 per cent of the original primary lowland and hill dipterocarp forests in S.E. Asia will have been logged and converted into forest estates with a much poorer species diversity, or clear-cut and transformed into monocultures of *Elaeis guineënsis, Hevea brasiliensis* or *Theobroma cacao.*

The problems. In many developing countries in S.E. Asia, the backbone of the economy is still largely dependent on the progress made in exploiting natural resources and in developing agricultural and wood-based industries. Therefore, any effort to conserve large tracts of primary forests is bound to meet an unsympathetic response or at best be considered a low priority by those responsible for managing and developing the country's economy. This is the case because in such a country, revenue and foreign exchange earnings derived from natural and semi-processed products exploited from the forests contribute very significantly towards economic development and the survival of the nation. This problem is further accentuated by the fact that in developing countries there is a great shortage of trained manpower who could carry out the necessary basic botanical and ecological research on those plants or groups of plants we wish to conserve. In consequence, resident scientists are more often than not unable to provide convincing arguments or accurate scientific data on, for example, how large an area has to be set aside from development to conserve a self-perpetuating breeding population of *Rafflesia hasseltii* or *Shorea leprosula.* Furthermore, in the developed countries industrialization and an advanced level of technological know-how have taken care of the economic needs of the population, and public awareness with regard to conservation has led to the protection of many natural habitats for public enjoyment, and the survival of species has reached a high level of priority; this receptive climate of public opinion does not yet exist in many developing countries.

Steps already taken or underway. Notwithstanding the problems mentioned above, conservation efforts in S.E. Asia do exist and are succeeding reasonably well. In Malaysia, for example, there are three well established National Parks (Taman Negara, Kinabalu and Mulu), which cover a total area of not less than 140,000 hectares. Apart from these, there are many conserved areas amounting to c. 450,000 hectares designated as Nature Reserves, National Nature Monuments, Wildlife Sanctuaries, Game Reserves, Virgin Jungle Reserves, Field and Research Stations, etc. Along with these reserved areas, there are also a number of organizations and societies concerned with conservation problems (e.g. the Malayan Nature Society, the Environment Society, the World Wildlife Fund). Similarly in Indonesia, the Philippines and Thailand, such activities also exist.

The various rôles of tropical botanic gardens in the conservation of rare and endangered plant genetic resources and as centres of research and education have been outlined and spelt out by many authors (Stewart, 1977; van Heel, 1977; Philipson, 1977; Schöser, 1977; Dransfield, 1977a; Soepadmo, 1977; etc.). From their accounts there seem to be at least five distinct functions which have been served by various botanic gardens in S.E. Asia. These are:

1. As a suitable place to grow collections of living plants.
2. As centres for basic botanical research.
3. As centres for introduction and propagation of plants of economic importance.
4. To provide a place for public recreation.
5. Education.

However, though at first glance the tropical botanic garden seems to have contributed substantially to the promotion of plant conservation, due to the shortage of trained manpower and facilities the impact of the research carried out so far is still far from adequate. It is normally the case that the research activities of botanic gardens are mainly concentrated on those plants which have direct and immediate economic importance (e.g. orchids, ornamental and medicinal plants). At the same time it is generally realized that since in S.E. Asia large-scale conservation of plant genetic resources *in situ* is out of the question at the moment and in the future, the only practical way to conserve as many plant species as possible, so that their complete erosion and extinction can be partly or wholly prevented, is by collecting, introducing and cultivating them in botanic gardens and arboreta. To be able to do this a lot more basic botanical and ecological research on their habitat requirements, reproductive and breeding systems, methods of cultivation and propagation, storage of propagating materials and economic potential, is urgently required.

With the help received from various international and national foundations and organizations, a modest start on this line of endeavour has been initiated in the region. For example, in Malaysia a new botanic garden of 35 hectares was established in 1974 within the campus of the University of Malaya in Kuala Lumpur. This garden is intended to supplement the already existing Water Fall Garden in Penang, the Ornamental Garden in Taiping (Perak), the Lake Garden in Kuala Lumpur and the Arboreta of the Forest Research Institute at Kepong, Kuching and Sandakan. The main purpose in establishing the new garden is first of all to initiate collection, introduction, propagation and research on selected species of wild fruit trees, ornamental plants and any others which have either scientific or economic interest. Secondly the garden is also intended as a teaching facility for university students; eventually, when it has been well established, it will be opened to the general public as well. In the four years since its establishment collections of seeds, seedlings or adult plants (in the case of herbaceous species) have been made from various threatened localities in the country. To date some 400 accessions have been made; these mainly consist of palms, aroids, gingers, orchids and fruit tree species. Likewise at the Water Fall Garden of Penang botanists of the University have been very active in collecting and propagating many species of ginger and epiphytic orchids. In the arboreta of the Forestry Department work on phenology and on vegetative propagation of many species of timber trees is also being actively conducted. In the neighbouring countries of Indonesia, the Philippines and Thailand similar activities are also being carried out. In Bogor (Indonesia) for example, a significant amount of research is being undertaken particularly on fruit trees, orchids, palms and vegetables (Dransfield, 1977a; Sastrapradja, 1975; Sastrapradja & Lubis, 1975; etc.). In addition a plan has been made to extend the present botanic garden at Bogor and to establish a laboratory for storing seeds.

CONCLUDING REMARKS

In conclusion, it may be re-emphasised here that:

1. The once luxurious, species-rich and complex tropical rain forests of S.E. Asia, in which thousands of plant species have existed since their origin and development in the geological past, are very fast disappearing.

2. Along with forest disappearance, those plant species of great scientific interest or having a great economic potential for future use will also go for ever.

3. Due to various unavoidable circumstances, large scale conservation of plant genetic resources *in situ* in S.E. Asia is completely impossible and impractical.

4. The only suitable place to conserve rare and threatened plant species is in botanic gardens or arboreta, National Parks and Virgin Jungle Reserves.

5. Within the limited time and space available, and in order to conserve as many plant species as possible, a lot more basic botanical and ecological research needs to be done.

6. Due mainly to the shortage of trained manpower, technological knowledge, funds and other supporting facilities, the amount of research and consequently also the impact of such activity on the conservation effort in general is not yet satisfactory.

7. To expedite and enhance the various efforts currently conducted in S.E. Asia to conserve rare and threatened plant genetic resources, external help in the form of training manpower and providing expertise through regular exchange of staff or by sending experts to work in tropical gardens for a reasonable length of time, is urgently required.

8. Since time is running out, unless a well co-ordinated effort by all concerned is forthcoming soon we may not be able to conserve many of the plant species now growing naturally in the rain forests of S.E. Asia.

REFERENCES

ASHTON, P. S. (1964). Ecological Studies in the Mixed Dipterocarp Forests of Brunei State. *Oxford Forestry Memoirs* **25**.

—— (1976). Factors affecting the development and conservation of tree genetic resources in S.E. Asia. *In 'Tropical Trees: Variation, Breeding and Conservation'* (eds J. Burley & B. T. Styles). Academic Press, London. Pp. 189–198.

—— (1977). A Contribution of Rain Forest Research to Evolutionary Theory. *Ann. Missouri Bot. Gard.* **64:** 694–705.

Boorsma, W. G. (1926). Notes about eastern medicine in Java. *Bull. Jard. Bot. Buitenzorg, 3,* **8:** 71–113.

Burgess, P. F. (1971). The Effect of Logging on Hill Dipterocarp Forests. *Malayan Nat. J.* **24:** 231–237.

Burkill, I. H. (1935, 1966). *A Dictionary of the Economic Products of the Malay Peninsula.* 2 vols.

—— (1951). The rise and decline of the greater yam in the service of man. *Advmt Sci. London* **7:** 443–448.

—— (1960). The organography and the evolution of Dioscoreaceae, the family of yams. *J. Linn. Soc. Bot.* **56:** 319–412.

Burley, J. & B. T. Styles (eds) (1976). *Tropical Trees: Variation, Breeding and Conservation.* Academic Press, London.

Chin, H. F. (1977). *Malaysian Flowers in Colour.* Tropical Press, Kuala Lumpur.

Cockburn, P. F. & A. J. Hepburn (1973). Agriculture and Forestry in the Development of Sabah. *In 'Proceedings of the Symposium on Biological Resources and National Development'* (eds E. Soepadmo & K. G. Singh). Malayan Nature Society, Kuala Lumpur. Pp. 39–43.

Coursey, D. G. (1972). The civilization of the yams: inter-relationships of man and yams in Africa and Indo-Pacific region. *Arch. & Phys. Anthrop. Oceana* **7:** 215–233.

Darus, Mohd. (1978). Forest resources of Peninsular Malaysia. *Malaysian Forester* **41:** 82–93.

Daryadi, L. (1978). Indonesian Forest Resources. *Malaysian Forester* **41:** 118–124.

Dongen, J. van (1913). *Beknopt Overzicht der meest gebruikte Geneesmiddelen in Nederlandsch Oost-Indie.* Koloniaal Instituut Amsterdam. Dieren.

Dransfield, J. (1977a). The Kebun Raya, Bogor, and the Conservation of Indonesian Palms. *In 'The Role and Goals of Tropical Botanic Gardens'* (ed. B. C. Stone). Penerbit Universiti Malaya, Kuala Lumpur. Pp. 181–185.

—— (1977b). Dryland Sago Palms. *In 'First International Sago Symposium: The Equatorial Swamp as a Natural Resource'* (ed. K. Tan). Kemajuan Kanji, Kuala Lumpur. Pp. 76–83.

Frankel, O. H. & E. Bennett (eds) (1970). *Genetic Resources in Plants: their Exploration and Conservation.* IBP Handbook No. 11. Blackwell Scientific Publications, Oxford.

Frankel, O. H. & J. G. Hawkes (eds) (1975). *Crop genetic resources for today and tomorrow.* International Biological Programme 2. Cambridge University Press.

Furtado, C. X. (1940). The Malayan Keladis and other edible Aroids. *M.A.H.A. Mag.* **10:** 11–17.

Heel, W. A. van (1977). The Importance of Tropical Botanic Gardens for

Morphological and Anatomical Research. *In 'The Role and Goals of Tropical Botanic Gardens'* (ed. B. C. Stone). Penerbit Universiti Malaya, Kuala Lumpur. Pp. 59–61.

HERKLOTS, G. A. C. (1972). *Vegetables in South-East Asia.* Allen & Unwin, London.

HEYNE, K. (1927). *De Nuttige Planten van Nederlandsch-Indië.* 3 vols. Buitenzorg.

HO, C.-C. (1971). The importance of Taman Negara as a gene pool of major cultivated plants. *Malayan Nat. J.* **24:** 215–221.

HOLTTUM, R. E. (1962). *Gardening in the lowlands of Malaya.* Strait Times Press, Singapore.

HOU, DING (1978). Florae Malesianae Praecursores LVI. Anacardiaceae. *Blumea* **24:** 1–41.

ISMAIL SAIDIN (1977). *Bunga-Bungaan Malaysia.* Dewan Bahasa & Pustaka, Kuala Lumpur.

JACOBS, M. (1974). Botanical Panorama of the Malesian Archipelago (vascular plants). *In 'Natural Resources of humid tropical Asia'.* UNESCO. Pp. 263–294.

JAMIL, MOHD. (1973). The Role of Agriculture in National Development. *In 'Proceedings of the Symposium on Biological Resources and National Development'* (eds E. Soepadmo & K. G. Singh). Malayan Nature Society, Kuala Lumpur. Pp. 5–15.

JARRETT, F. M. (1959). Studies in *Artocarpus* and Allied Genera, III. A Revision of *Artocarpus* Subgenus *Artocarpus. J. Arnold Arbor.* **40:** 113–155.

—— (1960). Studies in *Artocarpus* and Allied Genera, IV. A Revision of *Artocarpus* Subgenus *Pseudojaca. J. Arnold Arbor.* **41:** 73–140.

JONG, K., B. C. STONE & E. SOEPADMO (1973). Malaysia Tropical Forest: An Underexploited Genetic Reservoir of Edible-fruit Tree Species. *In 'Proceedings of the Symposium on Biological Resources and National Development'* (eds E. Soepadmo & K. G. Singh). Malayan Nature Society, Kuala Lumpur. Pp. 113–121.

KIEW, R. (1977). Taxonomy, Ecology and Biology of Sago Palms in Malaya and Sarawak. *In 'First International Sago Symposium: The Equatorial Swamp as a Natural Resource'* (ed. K. Tan). Kemajuan Kanji, Kuala Lumpur. Pp. 147–154.

KOCHUMMEN, K. M. (1973). Lesser Known Timber Trees of Malaysia. *In 'Proceedings of the Symposium on Biological Resources and National Development'* (eds E. Soepadmo & K. G. Singh). Malayan Nature Society, Kuala Lumpur. Pp. 123–129.

KOSTERMANS, A. J. G. H. (1958). The genus *Durio* Adans. (Bombac.). *Reinwardtia* **4:** 357–460.

—— (1966). A monograph of *Aglaia,* sect. *Lansium* Kosterm. (Meliaceae). *Reinwardtia* **7:** 221–282.

LEE, P. C. (1973). Multi-use Management of West Malaysia's Forest Resources. *In 'Proceedings of the Symposium on Biological Resources and National Development'* (eds E. Soepadmo & K. G. Singh). Malayan Nature Society, Kuala Lumpur. Pp. 93–101.

LEE, Y. H. & Y. P. CHU (1974). *Commercial Timbers of Peninsular Malaysia.* Department of Forestry and Malaysian Timber Board, Kuala Lumpur.

Low, C. L. (1975). Fruits in Peninsular Malaysia. *In 'S.E. Asian Plant Genetic Resources'* (eds J. T. Williams, C. H. Lamoureux & N. Wulijarni-Soetjipto). BIOTROP, Bogor, Indonesia. Pp. 47–52.

MARTIN, F. W. (1975). Yams of S.E. Asia and their future. *Loc. cit.* Pp. 83–96.

MEIJER, W. (1969). Fruit Trees in Sabah (North Borneo). *Malayan Forester* **32:** 252–265.

NALAMPOON; A. (1978). Forest Resources in Thailand. *Malaysian Forester* **41:** 114–117.

OCHSE, J. J. & R. C. BAKHUIZEN VAN DEN BRINK (1931). *Vegetables of the Dutch East Indies.* Archipel Drukkerij, Buitenzorg, Java.

PHILIPSON, W. R. (1977). The Importance of Living Collection for Studies of Phylogeny within the Dicotyledons. *In 'The Role and Goals of Tropical Botanic Gardens'* (ed. B. C. Stone). Penerbit Universiti Malaya, Kuala Lumpur. Pp. 69–72.

POORE, M. E. D. (1968). Studies in Malaysian rain forest. I. The forest on Triassic sediments in Jengka Forest Reserve. *J. Ecol.* **56:** 143–196.

PRANCE, G. T. (1977). Floristic Inventory of the Tropics: Where Do We Stand? *Ann. Missouri Bot. Gard.* **64:** 659–684.

PRINGLE, S. L. (1969). World supply and demand of hard-woods. *Proc. Conf. Trop. Hardwoods.* Syracuse.

PURSEGLOVE, J. W. (1972). *Tropical Crops: Monocotyledons.* Longman, London. 2 vols.

PUSPARAJAH, E. & H. Y. CHAN (1973). Optimising of land use for perennial crops in W. Malaysia. *In 'Proc. Symp. Nat. Util. Land Res. Malaysia'.* Universiti Pertanian Malaysia. Pp. 7–22.

REYES, M. R. (1978). Philippine Forest Resources. *Malaysian Forester* **41:** 104–113.

RICHARDS, P. W. (1952). *The Tropical Rain Forest.* Cambridge University Press.

SALLEH, MOHD. NOR (1973). Forest resources for economic development and conservation. *In 'Proc. Symp. Nat. Util. Land Res. Malaysia'.* Universiti Pertanian Malaysia. Pp. 53–62.

—— & H. T. TANG (1973). Some Aspects of the Utilization and Conservation of the Forest Resources of West Malaysia. *In 'Proceedings of the Symposium on Biological Resources and National Development'* (eds E. Soepadmo & K. G. Singh). Malayan Nature Society, Kuala Lumpur. Pp. 103–111.

SASTRAPRADJA, S. (1975). Tropical fruit germplasms in S.E. Asia. *In 'S.E. Asian Plant Genetic Resources'* (eds J. T. Williams, C. H. Lamoureux & N. Wulijarni-Soetjipto). BIOTROP, Bogor, Indonesia. Pp. 33–46.

—— & K. KARTAWINATA (1975). Leafy vegetables in the Sundanese diet. *Loc. cit.* Pp. 166–170.

—— & S. H. A. LUBIS (1975). *Psophocarpus tetragonolobus* as a minor garden vegetable in Java. *Loc. cit.* Pp. 147–151.

SCHÖSER, G. (1977). The Conservation of Tropical Orchids. *In 'The Role and Goals of Tropical Botanic Gardens'* (ed. B. C. Stone). Penerbit Universiti Malaya, Kuala Lumpur. Pp. 175–179.

SOEGENG-REKSODIHARDJO, W. (1962). The species of *Durio* with edible fruits. *Econ. Bot.* **16:** 270–282.

SOEJARTO, DJAJA D. (1965). *Baccaurea* and its use. *Bot. Mus. Leafl. Harvard Univ.* **21**: 65–104.

SOEPADMO, E. (1977). Conservation of Wild Fruit Tree Species. *In 'The Role and Goals of Tropical Botanic Gardens'* (ed. B. C. Stone). Penerbit Universiti Malaya, Kuala Lumpur. Pp. 207–210.

—— (in press). Genetic resources of Malaysian fruit-trees. *In 'Proceedings SABRAO-Malaysia Symposium on Genetic Resources of Plants, Animals and Microorganisms'.*

—— & T. KIRA (1977). Contribution of the IBP-PT research project to the understanding of Malaysian forest ecology. *In 'A New Era in Malaysian Forestry'* (eds Sastry, Srivastava & Manap). Universiti Pertanian Malaysia. Pp. 63–90.

SOEPARDI, S. (1964). *Apotik Hijau.* P.T. Purna Warna, Surakarta, Indonesia.

STEENIS, C. G. G. J. VAN (1951). The delimitation of Malaysia and its main plant geographical divisions. *Flora Malesiana, 1,* **1**: lxx-lxxv.

—— (1971). Plant conservation in Malaysia. *Bull. Jard. Bot. Nation. Belg.* **41**: 189–202.

STEENIS-KRUSEMAN, M.J. VAN (1953). Selected Indonesian medicinal plants. *Bull. Org. Sci. Res. Indon.* **18.**

STEWART, W. S. (1977). Research and Education Programmes for a Tropical Botanic Garden. *In 'The Role and Goals of Tropical Botanic Gardens'* (ed. B. C. Stone). Pernerbit Universiti Malaya, Kuala Lumpur. Pp. 55–58.

VALMAYOR, R. V. & R. C. ESPINO (1975). Germplasm resources for horticultural breeding in the Philippines. *In 'S.E. Asian Plant Genetic Resources'* (eds J. T. Williams, C. H. Lamoureux & N. Wulijarni-Soetjipto). BIOTROP, Bogor, Indonesia. Pp. 56–76.

VAVILOV, N. I. (1951). The origin, variation, immunity and breeding of cultivated plants. *Chron. Bot.* **13**: 1–366.

WHITMORE, T. C. (1971). Wild fruit trees and some trees of pharmacological potential in the rain forest of Ulu Kelantan. *Malayan Nat. J.* **24**: 222–224.

—— (1973a). Frequency and habitat of tree species in the rain forest of Ulu Kelantan. *Gard. Bull. Sing.* **26**: 195–210.

—— (1973b). *Palms of Malaya.* Oxford University Press, Kuala Lumpur.

—— (1975). *Tropical rain forests of the Far East.* Clarendon Press, Oxford.

WONG, I. F. T. (1971). *The present land use of W. Malaysia 1966.* Div. Agric., Ministry of Agriculture and Lands, Kuala Lumpur.

WYATT-SMITH, J. (1966). Ecological Studies on Malayan forests. *Research Pamphlet No.* **52.** Forest Research Institute, Kepong, Malaya.

The Collection, Establishment and Distribution of Natural Source Plant Material

J. B. Simmons

Royal Botanic Gardens, Kew, England

With the relentlessly increasing threat to the world's flora apparently reducing truly natural areas in every continent to the level of fragile reserves, it is sensible to apply logical and well defined procedures as a basic approach to plant collecting, accepting that where critically endangered species are concerned no effort should be spared to ensure their survival.

In the three years that have elapsed since the last Kew Conservation Conference, progress has been made in implementing many of the ideas promulgated through the Agreed Resolutions. But this is only a beginning and in proportion to the current scale of destruction, particularly to both tropical and subtropical floras, the amount of funds available to plant conservation on a world scale is derisory. The time is now opportune to open new avenues of work.

In this respect introduction into botanic gardens and distribution of plant material of known natural source is of great importance. Others cannot be expected to take conservation seriously if botanic gardens do not show this intention in their everyday work.

BACKGROUND

The historic pattern of plant collecting associated with the development of western and western-influenced botanic gardens is well documented from its early beginnings to a peak closing in the first four decades of this century, when the influx of new plants, particularly those from China, had a profound effect on the form of our gardens. After the war years this was followed by a period extending through the 1950s and into the 1960s when the exchange of material between collections became the dominant method of plant acquisition by botanic gardens. But the developing requirements of the newer plant sciences associated with taxonomy for plant material of known origin, and the disrepute into which exchange had fallen through offerings of unauthenticated material of dubious antecedence, has led through the last decade to a return by botanic gardens to an increasing involvement in the direct acquisition of natural source material.

Generally, the pattern of collecting trips has been rather random, slightly opportunist, in that it has been possible to attach botanical collectors to expeditions going to remote regions for other purposes. Significant for the Royal Botanic Gardens at Kew has been the establishment over the last six years of

75

an expedition programme which has combined botanical and horticultural objectives. Most of the botanical work at Kew is undertaken on plants from the tropics and subtropics, and so the collections tend to be well supplied both through the resultant contacts and the attachment of horticulturists to Kew's botanical expeditions. For hardy plants separate expeditions are arranged by the Living Collections Division.

There is a tendency among botanic gardens for those holding collections of native plants to have local field collecting programmes, and large gardens with international collections often collect overseas. However there are a great many gardens that do not have collecting programmes, and who therefore have to acquire material by exchange or even, as in the UK, by purchase from commercial sources. One consequence for those gardens with collecting programmes is therefore the further requirement to act as suppliers of plant material to other botanic gardens. As far as threatened plants are concerned, the more widespread the species is in cultivation the more secure is its status.

While acknowledging the need for a selective input of plant material as the basis for a botanic garden's work, there are many parameters to this concept that should be observed. Long gone are the days of exploitive 'discovery collecting'. Often, even in remote regions, the main floral components are well known and collected, and a more profitable approach is through detailed specialist collecting on a taxonomic or regional basis. From the earlier 'find and take' approach, there is a movement to a more defined position where recording, study and conservation considerations are taken into account.

Where overseas collecting is involved the charge of scientific imperialism should never arise. It is fundamental to conservation aims that all overseas expeditions are linked through the host countries' scientific institutions to local botanical institutes or universities. In this way duplication of expeditions can be avoided and there can be a real exchange of knowledge and support, so helping to encourage research and conservation work in other countries.

Local advice on the distribution and frequency of species can further help with the recognition of rarities to avoid over-collection. Recording such information is also important so that the rarity status of plants becomes known. For example only relatively recently a previously unsuspected stand of *Camellia granthamiana* was found on Tai Mo Shan in the New Territories of Hong Kong; we now know that what was believed to be extremely rare has a more secure, if still vulnerable, status in nature. In contrast *Broussonetia kazinoki*, which is relatively uncommon in cultivation, is relatively frequent in its natural habitat of Korea, Japan and mainland China, where I have for example seen it growing in the Lushan region of Kiangsi Province.

All expeditions should have a clear purpose and any collecting of plants should be set against known requirements. Very often it is necessary to collect living material for scientific study purposes, and much of the material received at Kew comes under this heading, with only a relatively small proportion used for amenity horticulture.

In an established garden only a relatively small amount of material is required from each new introduction. Even so, from each batch of seed or seedlings, surplus arises which can be offered to other gardens or institutes. With the current emphasis on natural source material such plants are generally welcomed by others, and additionally, if the plants are raised from seed, a wider genetic sample of the material becomes established in cultivation.

COLLECTING

Where material is collected by those inexperienced in horticultural techniques a high failure rate may be expected with unsuitable material arriving in very poor condition. The following notes are given as a brief guide to the means of improving this situation.

Timing. In the tropical rain forest collecting may be undertaken throughout the year, but since seed production is not seasonal, as in temperate regions, most accessions from tropical regions tend to be in the form of plants or vegetative propagules. In those seasonal parts of the tropics where dry periods occur, collecting should be after the rainy season when plants can be located and identified. Obviously, as in desert regions, perennating plants can be more readily transplanted when dormant, and collecting may be delayed until the seasonal growth cycle is almost completed. Bulbs, for example, may well not be found again if left too late.

In cool temperate regions late summer yields the seeds and perennating organs of herbs whereas for trees autumn provides the main seed harvest. The problem with generalizations, however, is that in biology there are so many variables: latitude, altitude, proximity of mountains and land mass all combine with climate to cause variations. Advance local advice is important in order to estimate the best timing for an expedition, particularly when in the search for more hardy variants, high altitude forms are required.

Equipment. The polythene bag with its moisture-retaining properties has gained universal use. In field operations two polythene sacks, one for empty bags, labels, etc., the other for plants in individual bags, ensures least damage to the material collected. Unused equipment should not be mixed with live plants.

Essential equipment includes field note books, tie-on plastic labels, waterproof ink pens, seed packets, knife, secateurs, a long-arm pruner for specimens from trees, a bulb trowel (an ice axe is a good alternative for digging up bulbs), a plant press and paper for voucher specimens, a camera for recording plant and location, seed packets, an altimeter, a map (for locations) and a compass, along with suitable cardboard boxes, polystyrene liners, tape, string and labels for packing and shipping out the plants by air. It is advisable to send such equipment out in advance but this is not always possible; shipping and Customs requirements for unaccompanied items can sometimes take months to clear.

Recording. This is an essential but often omitted aspect of collecting living specimens. Caring for the plants takes a lot of time and full documentation under field conditions means working late into the evening. The requirements of full field records have been described elsewhere and it is hoped to publish a guide to field collecting (which will include these details) as a follow-up to this conference. A copy of the field notes giving the minimum requirement of identification, location and edaphic notes should accompany the living material when it is despatched. As already indicated it is advisable with important specimens to supplement the written notes by the addition of a voucher specimen and photograph.

SELECTION OF MATERIAL

Whenever a rare species is to be established in cultivation, it is important to use the minimum amount of material consistent with genetic requirements. Firstly this ensures that the wild population is minimally reduced; secondly it makes economic sense to send small propagules; and thirdly small juvenile individuals are easier to establish.

Seed, when available, is obviously the best method of plant introduction, particularly since it ensures genetically varied individuals in cultivation, but as with other parts of plants it is not just a question of 'packeting and forgetting', particularly for fleshy seeds. Many seeds, such as of *Papaver* species, are naturally dry and small, and can be stored readily in small paper envelopes. Seeds of many trees and shrubs, however, particularly those possessing a large endosperm or formed as a fleshy fruit, will need cleaning and drying in the field, while others will require keeping moist but not so moist that they decompose. Some fleshy fruits need cleaning fairly quickly; many cucurbits are a case in point, though their seed usually stores well under dry conditions. Cleaning the fleshy fruits of *Crataegus* can usually be delayed for a month or so until they are brought back to the home institute.

The tropical South American *Mauritia* palms are difficult to cultivate under glass. They grow naturally by stream-sides or on swamp islands, and the ripe seed will fall naturally into deep moist leaf litter, where their hard scaly seed coat slowly decomposes. Collected and held in moist leaf litter in a polythene bag, the large golf-ball sized seeds of *Mauritia setigera* that I collected in Trinidad survived to germinate at Kew. The seed of *Quercus* are noted for their storage problems, as they germinate very rapidly when taken from a cold store at 9°C. They also usually germinate naturally in moist leaf litter, but germination in transit need not be a problem if the seeds are well packed. I collected in May, after germination had commenced, seeds of a large buttressed oak in montane forest in the south of Hainan, and despite being kept for 4–5 weeks in a polythene bag with leaf litter, they all survived to continue their germination at Kew.

The seed of many tropical and subtropical plants are short-lived, but rapid air transport can overcome this problem, as for example with the endangered *Camellia crapnelliana*. For *Cinnamomum*, however, even the most rapid transport may not be quick enough. *Phenakospermum* from Brazil and Guyana, like its close relative *Ravenala*, also presents difficulties but there may be other factors since even good seed usually gives a very low percentage germination. However these are all extreme examples and for most plants, given that the seed is kept in good condition, air transport represents the easiest method of plant introduction.

Ferns may be collected as spores although sporelings represent a more certain method. Fertile fronds, taken as vouchers, yield a great quantity of spores when dried. As with seeds, the germination of spores is variable; *Gleichenia* and its relatives tend to have short-lived spores and *Marattia* and *Angiopteris* can only be grown from spores when specialized aseptic techniques are available to raise the prothalli. Where growing seasons apply, the equivalent of late summer or autumn is the right time to seek fertile fronds; this is notable with tree ferns such as *Cyathea* where the early season fronds are all sterile. Care must be exercised to avoid cross contamination between packets of spores.

If seed is not fully ripe it is still probably worth taking a capsule, especially for orchids. There is sometimes an advantage in sowing fully formed but not fully ripe seed in that the natural chemical dormancy controls may not yet have been laid down.

Seedlings.   Overall, I have found this to be one of the most successful methods of direct introduction from the field, particularly for woody species and for tropical plants. Seedlings (or sporelings) in their first year or so of growth, preferably growing in leaf litter or other loose soil so that the entire root system can be taken up without damage, are the best subjects. The roots may be wrapped in moist moss or absorbent tissue paper and the seedlings kept in a loosely fastened polythene bag. With some care, checking daily to see that they are not too wet or dry, and stored in a cool shaded place, a high rate of success can be achieved. I have for example applied this technique to plants as diverse as *Cedrus libani*, *Laguncularia racemosa* (a mangrove from the West Indies) and two Oleaceae, *Chionanthus macrophyllus* and *Olea dioica* from the forests of Hainan. The latter two survived five weeks transit when, in comparision, cuttings of two jasmines (also Oleaceae) had died within ten days.

If the seedlings are received in the propagation house within a few days of collection then they can be pricked out and held in a closed frame until recovered, but if held for longer, and also if at all desiccated, it is better to treat them as cuttings by inserting them in a propagation bed with controlled heating and overhead intermittent mist until new roots have developed.

The advantage of seedlings is that they can be taken in any season, and only a small quantity are required. Where they are plentiful thinning out from a natural bed of seedlings is unlikely to reduce the population since only a few would normally achieve maturity.

Cuttings.   It is quite difficult to introduce plants from cutting material. There are however some exceptions, such as *Salix aegyptiaca* which I found survived easily as hardwood cuttings when collected in the Caspian forest in autumn. Selecting the right material is a problem, as is keeping the plants alive until they reach the propagation unit. For trees and shrubs timing is also very important if the right stage of semi-mature material is to be found. It is usually preferable to prepare the cuttings at the time they are taken; this should include trimming off all surplus leaves.

One possibility is to take hardwood material (in 20 cm lengths), which will survive transit and with luck can be forced into growth in the propagation house, so that the resultant young shoots can be removed and rooted as cuttings. Once when trying to bring back *Findlaya*, a climbing Ericaceae from Trinidad, I decided to take small branches and keep their bases in a container of water; the whole branches were enclosed in a polythene tent for about a week until the day of despatch, when semi-ripe tip cuttings were removed and wrapped for despatch. Happily they rooted readily at Kew.

Aside from trees and shrubs, the herbs are often a better prospect as cuttings, particularly Acanthaceae for example. Trailing or climbing plants which have formed roots along their stem provide the best opportunity; hoyas, for example, can be readily propagated by this means as can *Trachelospermum* and some of the climbing aroids. Indeed any cuttings with roots attached have a good chance of survival and this applies particularly to suckers—as may be found on *Clerodendrum*.

It is difficult to make general recommendations for handling cuttings, particularly of trees and shrubs, since much depends on the condition of the material, and the time that elapses between taking the cutting and its arrival in the propagation bed. A loosely tied polythene bag or rigid PVC container is the usual choice of holding receptacle, and depending on the subject, it may be desirable to wrap the cutting bases in damp moss—though this can cause rotting in some subjects.

Succulent plants such as cacti and the succulent euphorbias and asclepiads (e.g. *Caralluma*), travel well in dry newspaper. With stem succulents, cutting material from the stem usually survives, the only danger being if it sweats and rots in transit. Aquatic plants such as *Nymphaea* travel best wrapped in wet newspaper held in an unsealed polythene bag.

Plants. Rather than taking whole plants, it is often possible to divide off a small part, particularly from those with a rosette habit such as many *Primula* spp. or perennating organs such as bulbs. Dormant bulbs are of course easily handled, but if in growth they may need to be held in an open polythene bag. Among corms, *Cyclamen* travel fairly well in growth. Division *in situ* can be hard work with some of the Gramineae and larger monocots such as *Yucca*. With the many zingibers and heliconias, for example, it is advisable to look for a specimen that has been cut back or damaged in some way in the hope of finding a young piece of rhizome with a small shoot. Most of the imported tropical orchids are received at Kew as pieces of rhizome; fortunately most of the epiphytic species tend to be fairly tough, and can survive in dry newspaper, though those from moist humid conditions and forest floor species may well prefer the benefits of a polythene bag. Similarly bromeliads are difficult to kill although some montane species from Venezuela and Trinidad appear susceptible to the warmth of the lower altitudes and may decay in transit.

Generally, ferns have to be kept in a closed polythene bag, and filmy ferns double wrapped, in a sealed polythene bag within another polythene bag.

In general with plants, the younger the material the better are the chances of success. Care should be taken not to damage the roots, and if possible some soil should be retained or the roots should be wrapped in moss or clean white tissue depending on the local plant health regulations for export.

The advantage of vegetative techniques is that they do not reduce the wild population. If the plant is particularly rare, then it is obviously essential to use the method of propagation that causes the minimum disturbance to the wild population.

Since expeditions to remote places are expensive and not likely to be repeated, different types of propagating material should be taken from the species concerned to help ensure success.

QUANTITY

The collecting requirements for seed banks have been devised to provide wide genetic sampling. For example, Hawkes* has recommended 50 capsules be taken from separate specimens over one square km. Where very rare plants are concerned, this sampling might not be practicable or desirable, and a small seed sample may be all that is available. For vegetative propagation of rarities, if the population is large enough, then it is advisable to try to achieve the desired samples from ten separate individuals, not clones, over one square km. Genetic sampling is vital when future re-introduction to the wild is planned, since the broader the genetic base and the more variable the seedlings, the better the chance of successful re-establishment. Also, with a good sample it is possible to establish small isolated 'breeding colonies' in cultivation, from which good seed can be produced. It is hoped to establish such a colony of native black poplars, *Populus nigra*, in the Queen's Cottage Grounds at Kew as this is now a very rare tree in Britain.

DISPATCH

Air transport has made it possible to send plants from one side of the world to the other with comparative ease. This is now so much taken for granted that the previous methods, such as sea shipment by Wardian Case, are all but forgotten. (The last Wardian Case to arrive at Kew came in 1962 from Fiji; it was affected by sea water and had been mislaid by British Rail between Liverpool and Kew.) It is possible to lose air freight consignments (as mentioned later), but this is not common.

In my experience it is advisable to leave all the plants upright and free in their polythene bags until the morning of shipment; then, if they are loosely tied, they can lose air at the final stage as they are compressed into the transport box—preferably a cardboard box lined with insulating inserts of polystyrene. String and strong tape are also needed at this point, along with clear labelling, since the box cannot be closed until after Customs inspection.

Advance telephone arrangements with the shipping agent should be made before leaving for the airport, complete with authorization (noting that the

* Circulated Paper: *Notes on the Conservation of Rare or Threatened Species in Botanic Gardens and Arboreta.* J. G. Hawkes, October 1976. Mimeo.

specimens are for scientific use) from the local host institute that should, Plant Health and Customs allowing, ensure the rapid return of material to the home base. If possible it is advisable to ship material out once a week, but specimens often have to be kept for much longer periods. In such cases a temporary nursery becomes vital.

Quarantine. When collecting in the field it is very advisable to select healthy material and to check it for pests and diseases. Unsuspected viruses will cause quarantine confiscation and cherished seed samples may well be full of beetles. To this end the development of *in vitro* exchange techniques, whereby clean material is raised and shipped in test tubes, offers great scope for exchange, but is not much use in field collecting.

ESTABLISHMENT

With rare plants the availability of good propagation facilities is important for their successful establishment in cultivation. Mist-propagation tends to be more effective than generally appreciated for the establishment of imported plants. Where plants have been shipped over a long distance they are usually desiccated, but worse their roots are often dead or beyond recovery. Thus the treatment for such plants needs to be as for cuttings to stimulate the production of new roots. Often plants are mistakenly potted on arrival; setting damaged and dead roots into a cold wet compost usually completes the plant's demise, whereas insertion under mist, with soil warming, helps to initiate new roots. With some subjects re-establishment may take many weeks, but once new roots are produced normal potting may commence.

The closed frame has a value in establishment since a humid atmosphere can be maintained, but there is also now commercially available a closed frame called a 'Dewpoint Cabinet', which is designed to assist in the establishment of imported material. It has in-built controlled heating and lighting, and provides a constantly moisture-saturated atmosphere, created by a small air pump which 'bubbles' air through water in the base of the cabinet. Its effectiveness has been well demonstrated by one example whereby cuttings of a new melastome were given to me in Caracas, with the request they be carried back to Kew. The cuttings were wrapped in damp newspaper and kept in a polythene bag. For over a week the cuttings were carried through hot and dry conditions, then shipped by air from Maracaibo with a consignment of other plants I had collected. The parcel was due to arrive in London the next day, but was lost by the airline and not found for one month. Not surprisingly some of the specimens were found to be dead on arrival. The worst affected plants, including the melastome, were then set in a 'Dewpoint Cabinet'. With careful attention over several weeks the melastome and several of the other specimens survived and eventually became established.

ENVIRONMENTAL INFORMATION

Just as understanding habitat requirements can assist in the successful cultivation of plants, so too can defining the needs of a plant in cultivation ultimately assist with re-introduction to the natural habitat.

Each habitat is a complex of climatic and edaphic considerations. Some plants have 'ecological flexibility'; an example is *Platanus orientalis* which can be found over a wide vertical amplitude (from 200 m below sea-level to 2,600 m) and range of soils, while other plants may have very narrow requirements. These 'flexible' plants are helpful to the horticulturist. We have for example happily grown *Glyptostrobus lineatus*, a deciduous conifer from southeast China, conventionally in a bed in the Temperate House and also had planted it out in one or two sheltered situations to see if it would prove hardy. What should not have escaped our attention, as I found when I saw the plant 'at home' in south China, is that it occupies exactly the same habitat as its American relative *Taxodium* and will grow happily in swamps and on the edge of pools.

The less tolerant subjects rarely provide cultivators with a second chance and in the 'really difficult' class *Heliamphora nutans* from Mt Roraima in Guyana comes to mind since, despite acquisition of much environmental knowledge on its habitat, it still cannot be maintained successfully in cultivation.

DISTRIBUTION

The number of specimens required to represent a common species in a botanic garden is relatively few. But where rare plants are concerned greater numbers give increased security and genetic diversity. Even then invariably more seed is received or seedlings raised than required and even if in small quantity the distribution of such natural source material proves a valuable exercise. The cost of the collecting and establishment of such material is high and in operational procedure at Kew the tendency now is to treat all natural source material as meriting extensive distribution; this has been applied over the last few years to material of known conservation interest. It is hoped that this approach of sharing material and thus sharing the responsibility for its maintenance and survival in cultivation will increase in future.

RE-INTRODUCTION

The new emphasis for botanic gardens to become involved in the conservation of their regional flora has led Kew to develop managed habitats such as woodland, meadow-land and wetland in its satellite garden at Wakehurst Place in Sussex, with a view to both maintaining the existing flora and also introducing and establishing some of the endangered species of southeast England. Such habitats are themselves often man-altered systems, the climax forest having been modified over many centuries by agriculture and forestry. Learning to manage these areas is a challenge but will be a significant step forward. Such ecosystems are a complex which we have to interpret and understand to guide us towards what must be tomorrow's challenge—the re-introduction of species into their original habitats.

Leaving aside the political and economic problems that currently constrain such work we must now accumulate the knowledge and experience to be ready to meet the challenge when opportunities for re-introduction occur.

Cultivation and Continued Holding of Aegean Endemics in an Artificial Environment

SVEN SNOGERUP

BCT Department, The Wallenberg Laboratory, Lund, Sweden

SUMMARY

The number, ecology and present status of the Aegean endemics are discussed. Objections are raised against the use of botanic gardens as a substitute for reserves. For continued holding of cultivated stocks, seed banking and production of new seed generations by experimental methods are suggested.

INTRODUCTION

I feel I represent here those taxonomists who are working biosystematically and I will present what may appear to be a rather negative attitude to plant conservation using the classical methods of botanic gardens. Also I will present my own views on the subject after considering the problems in more depth and in the light of some recent positive experiences I have had.

The Aegean endemics represent one example from the long series of geographical groups of endemics, all of which may be equally interesting. Seed samples or in some cases living plants are available for many species at our institute and at several other places. When the taxonomic work on the different groups has been finished, the question arises what should we do with the remaining seed, and whether it is at all worthwhile keeping any. If we come to particular conclusions concerning the Aegean endemics, these may apply also to several other floras or plant groups.

First we should consider some basic facts about the Aegean endemics such as their number, their ecology, their present status and their possible importance for different purposes. The area discussed includes the islands and peninsulas of the Aegean sea between Greece in the west, Turkey in the east and Crete in the south. Crete itself is not included, as it presents problems of its own and has another collection of endemics of roughly equal numbers, but Aegean species also occurring in Crete are included.

I have just started to make a report about the status and possible conservation measures needed for the Aegean endemics, for the Threatened Plants Committee of IUCN. The initial IUCN list contained 104 names, from which only a few can be excluded as misplaced. Several categories have to be added, such as the endemics of the island of Euboea, those having subspecific rank and those not

85

accepted in *Flora Europaea*. Thus the true figure will not be less than 125, and will probably be closer to 150. Some listed taxa are admittedly of questionable rank, but such problematical cases are often especially interesting and worthy of protection.

Most of the endemics are only found in small areas, within ranges of 10 to 100 km, but still with several populations and with a considerable number of individuals. Some, however, are very restricted point endemics, in a few cases with a few hundred individuals only. Concerning their ecology, 54 can be classed without hesitation as more or less obligate chasmophytes, three as occurring mainly at cliff bases, seven in rivulet courses and similar mesic habitats, four in different types of sea shore, two in the highest mountain areas, one in montane scree, one in forest, two in small islets and 15 in the more or less dry, common habitats of the phrygana or garigue type. These figures cover only some of the species, information on the others being at present insufficient, but they give a good idea of the proportions. Much the largest category is the chasmophytes, and among them are also most of the very local and rare species as well as those that are the most taxonomically isolated. It is admittedly possible that this dominance of the chasmophytes among the endemics reflects an ancient, much richer flora with destruction of several species in the more accessible habitats. But here we have to deal with the proportions as they are now and without speculation about the past.

THE PURPOSE OF KEEPING SPECIES IN CULTIVATION

What can be the purpose of keeping these endemics in continued cultivation, through seed banking or direct cultivation or both? I think three main purposes have to be taken into account:

(a) To save them from extinction.
(b) To keep material of them available for future scientific work.
(c) To keep material available for future plant breeding purposes.

Saving them from extinction is not necessary at present for most of the Aegean endemics, especially not for the chasmophytes which occur in rather inaccessible, self-protecting localities. Individual cliffs can of course be destroyed by mining, by building, by dumping of rubbish, or by similar activities, or even by the introduction of aggressive species like *Opuntia ficus-indica* and *Nicotiana glauca*. Most species have, however, a number of different populations, so that the destruction of one or a few will not necessarily be catastrophic. In the long run, however, some cliff reserves must be set aside for full protection of this rich and interesting chasmophytic flora. We now have enough detailed knowledge of the distributions to make a survey and point out about a dozen localities which may be called major cliff refugia. These rich localities will each contain several endemics, and as a rule also include the especially vulnerable point endemics. Some people are pessimistic about the possibilities of creating such reserves, but I am optimistic. I think it is possible to raise among politicians a national or local pride in these unique, scientifically interesting, local species. The cliff localities are also of very little commercial value and thus may be easy to set aside.

Maintaining living collections of plants for scientific purposes would for a variety of reasons be valuable for many of the endemics. Often much work has been expended to collect them, not only in tiresome and expensive travel but also often in hours of climbing, stone-throwing, etc. It would be good if this did not have to be repeated for new investigations, and the amount of collecting from the native populations would also be reduced. But for detailed investigations, intended to throw new light on some problems, a fully representative collection of pure material is needed and anything else will be of little value.

Keeping plants for future breeding purposes represents another problem. Only particular groups of plants are involved, but from these all populations should be kept as should perhaps be spare seed from crossings, etc. Even if such groups do not contain endemic species, their local variants will be of special interest and worth keeping as an easily accessible gene pool for different breeding purposes. In these cases the danger of repeated collections on a comparatively large scale should also be kept in mind.

THE DIFFICULTIES WITH CONTINUED HOLDING OF LIVE PLANTS

There are three main difficulties for any project to conserve material of the endemics in cultivation. The first concerns the difficulties of maintaining the plant itself and its correct labelling in a garden. We have had no difficulties in growing these species in an experimental greenhouse, but the situation is quite different in a conventional botanic garden. We have repeatedly delivered plants to an associated garden, only to see how sooner or later they disappeared or became mis-labelled during weeding or by exchange of labels. We also received many samples from other botanic gardens through seed exchange, and many of them turn out to be quite different to what was stated on the label. Then one must doubt the origin and purity of the rest. To carry out any large experimental work with such uncertain material cannot be anything but a waste of time. I consider that the continued keeping of such plants must be done in specialized cultivation sites, separate from the ornamental arrangements, to give any chance of success.

The second difficulty is to prevent crossings and thus keep the material pure. The size of the problem can be illustrated with an example from one of my earlier study groups, *Erysimum* sect. *Cheiranthus*. This group contains in the Aegean several taxa of questionable rank, some of them rather local. As an example take *E. naxense*, endemic to the Cycladian island of Naxos, and *E. senoneri* subsp. *icaricum* of the island of Ikaria. *Erysimum naxense* is a large plant with densely placed large leaves, a large apical inflorescence with many-flowered partial inflorescences having siliquas turned to one side. *Erysimum senoneri* subsp. *icaricum* is a small, small-leaved plant, branching and producing several few-flowered inflorescences with siliquas in all directions. They give an impression of being rather distantly related when seen together and I do not hesitate about their taxonomic treatment. However their F_1 hybrid is a large, vigorous plant which has almost full pollen fertility and sets seed readily both on crossing and self-pollination. The F_2 and F_3 generations admittedly suffer a lot of breakdown, but some lines maintain vigour and fertility. Even if these secondary hybrids would have little chance in nature, they can easily be kept in cultivation. If these two plants or another combination of materials from the group are cultivated close to each other, as is normally the case in a garden, crossings will take place very easily. The crossings are also favoured by a partial self-sterility of these plants which secures a high percentage of out-breeding. Collecting of seeds from such plants after free flowering will have no chance of giving acceptable results. In the case of *Erysimum* even species of other sections with other basic chromosome numbers will hybridize with species of section *Cheiranthus* and sometimes give partly fertile progeny. The same pattern of interfertility and partial self-sterility is found in a whole series of plants from both the Cruciferae and other families. There are only two possible ways to avoid such hybridization. One is to grow the actual plants absolutely isolated from all relatives, preferably in an insect-tight greenhouse on their own. The other is to maintain breeding isolation and prevent crossings between populations by the same methods as during experimental work.

The third difficulty with continued cultivation is to retain the variability and the internal genetic system of the material unchanged. This can again be illustrated with *Erysimum*. The populations are small, about two thirds of them below 100 individuals. From them, we should suppose that each population becomes rapidly homozygous, turning into what Greuter called a "state of non-evolution", easy for an old-fashioned taxonomist to handle and without interesting problems. For the *Erysimum* material I have, however, shown that considerable variation and heterozygosity is retained for leaf forms and petal forms, and therefore presumably also for other characters (Snogerup, 1967). Now if these plants do not behave as should be expected from their population sizes, then responses must depend on their own internal genetic properties. In the case of the *Erysimum* I am quite sure that considerable mutation rates are at work; in other cases probably balanced structural heterozygosity as proposed by Runemark & Heneen (1962) occurs. To retain the full value of this material for future investigation with improved methods, the population should be treated so that both its genetic constitution and the original variability is preserved unchanged. In nature characters are preserved in an equilibrium between the internal properties of the plant, the genetic drift effects and the strong selection of the cliff habitat. To simulate all this in a botanic garden seems impossible, especially as the factors vary from one population to another.

If we admit that we cannot do *the* best, i.e. simulating fully the natural conditions, then we can perhaps still do *our* best by keeping the material, at least for short intervals of time. I think the only acceptable solution is:

(a) To keep the number of artificial generations as low as possible by seed banking for long intervals.

(b) To raise the seed for a new generation by means of controlled crossings between the individuals of each population separately. This should be done so as to include individuals covering the full variation in morphology as well as in fertility and other observed properties. For this purpose the material has to be observed morphologically and related to fertility, chromosome number, etc.

How much work does the proposed method lead to? Say that a certain garden is preserving 500 endemics, with 10 separate populations of each. If the acceptable banking period is 50 years, then some 100 populations will be taken into cultivation every year. To make the necessary controls and crossings would then, according to my experience from experimental work, require at least one person with good biosystematic experience, one laboratory assistant and one specialized gardener. They will also need a large special greenhouse where the work can be done free from rain, wind and the plants protected from handling by visitors. This means that if a garden wants to keep such material for future scientific purposes, it must obtain resources for these special staff and greenhouses and also be prepared to change from the traditional practice to new methods. Then there can of course be no objection against removing portions of seeds for cultivation and exhibition in the open parts of the garden. They will be destroyed, but they can be replaced every time a new controlled generation is raised.

THE ATTITUDES OF BOTANISTS

On completion of several earlier works with Aegean endemics no seeds for continued storage were produced, as no one knew where they could be deposited. And the remaining seed samples have been either incorporated with herbarium sheets or stored in various not very elaborate ways. But if a permanent organization for the maintenance of such material is established, then I think most botanists are prepared to change their practice and also include the preparation of seed reserves as a routine part of their work. Already new institutions for continued storage of such material have been created or are planned, and this I think has already influenced our attitudes. One of our main groups on which we are working at present is the *Brassica oleracea* group, including also many variants from outside the Aegean. From this material we are producing a reserve of seeds, both from the natural populations and from the hybrids made. The final keeping of this seed has not been solved, but the development of seed banks for different purposes will certainly open possibilities before the work is finished. This material also illustrates, in my opinion, that it is not only the endemics that are of interest but also a lot of other plants which are of interest from different angles. Even if the species as a whole is not at all threatened, it may also in some cases be desirable to save the local populations.

CONCLUSIONS

Most of the Aegean endemics are not directly threatened at present, but are vulnerable to more radical future changes in the environment. Saving them through permanent keeping in a botanic garden will mean only a saving of the species in its taxonomic sense from total extinction. Its genetic content as well as its genetic system will in most cases be seriously changed after the transfer to an artificial environment. Therefore saving of species should be carried out by the setting up of reserves whenever possible. The keeping in cultivation is just a last-resort substitute.

Seeds for further investigation may with modern banking techniques be kept for rather long periods without any intermediate generations, for some materials around 50 to 100 years. If one or a few generations are grown on at such intervals, then the time can be much prolonged.

Those taxonomic groups containing or being closely related to cultivated plants should be given special attention. From them larger and more diverse material should be kept, even if this means extra work and more space is needed.

All materials kept for conservation purposes should not be included among the routine exhibition plants of a botanic garden. They should instead be kept under special control. The number of generations should be kept as low as possible by banking of seeds for long intervals. New generations of seeds should be raised by controlled crossings to ensure, as far as possible, an unchanged genetic constitution.

REFERENCES

Snogerup, S. (1967). Studies in the Aegean Flora IX. *Erysimum* sect. *Cheiranthus*. B. Variation and evolution in the small-population system. *Op. Bot.* **14:** 1–86.

Runemark, H. & W. Heneen (1962). Chromosomal polymorphism and morphological diversity in *Elymus rechingeri*. *Hereditas* **48:** 545–564.

Multiplication and Re-introduction of Threatened Species of the Littoral Dunes in Mediterranean France

LOUIS OLIVIER

Parc National de Port-Cros, Hyères, France

International inventories that have been made in recent years show that many flowering plants throughout the world are likely to become extinct. In the case of numerous species, it is at present difficult to be confident of their survival under strictly natural conditions. In most cases it will be necessary to provide for their multiplication in special organizations.

The justification for this work lies not only in the fact that these species are interesting from an historical or scientific point of view, but more especially in the fact that they are potentially useful to man. For in order to carry on his various activities and meet his needs in terms of energy, raw materials, food, medicines, etc., man has to turn—and will increasingly have to turn—to those inexpensive natural resources whose use does not threaten his own survival.

The various dangers are all the more serious in the Mediterranean region owing to the fact that here special conditions of climate and soil have produced some very rare species. In fact, for every five species native to Europe, four are of Mediterranean origin.

At the same time, this very vulnerable area, in which the consequences of uncontrolled exploitation can bring about truly catastrophic changes in the flora and fauna, is being seriously and continuously damaged by fire, over-grazing, disorganized town-growth or thoughtless development in the interests of tourism.

In the French Mediterranean region, the coastal areas have been particularly affected over the last twenty years. It is becoming more and more difficult to find dunes and sand-bars in an undisturbed state. In 1972 a Ministry of the Environment report listed only five sites where these biotopes remained intact. Two were in the *Département* Hérault, one in the *Département* Aude, one in the *Département* Bouches-du-Rhône and one was in Corsica. At all other points along the coast, these habitats had already become badly deteriorated.

This state of affairs suggests two lines of thought. First of all, it is noticeable that species which in the past were fairly common are becoming more and more rare because of the disappearance of the very special biotopes to which they are strictly confined. Hence in the Mediterranean region, more than elsewhere, protection should cover not merely endemic plants but also those species which have special ecological requirements.

Secondly, efforts should be made to preserve those sites where the right environment can be maintained or reconstituted, for after all raising plants in a botanic garden is not enough to preserve the whole gene pool of a species. The

91

multiplication of small populations carries with it the well-known risk of genetic drift. It is true that gene banks are no substitute, but they should be considered as an indispensable complement to the protection of natural habitats.

It is with these considerations in mind that an experiment has been launched for the reconstitution and safeguarding of a dune several kilometres long in the vicinity of Hyères. Different parts of this dune afforded examples of the three facies of this type of formation as described by K. Lordat, namely:

1. Young dunes with *Agropyron junceum* and *Sporobolus pungens*, in which are found the following rare and uncommon species, *Limoniastrum mono-petalum*, *Diotis candidissima (= Otanthus maritimus)*, *Matthiola sinuata* and *M. tricuspidata*, for which this dune is the only locality on the French mainland.

2. Mature dunes with *Ammophila arenaria*, *Echinophora spinosa* together with *Medicago marina* and *Eryngium maritimum*, which have become rare.

3. Senescent dunes with *Crucianella maritima*, *Artemisia glutinosa* and *Pan-cratium maritimum*, on which, in particular, plants less typical of sand-dune floras such as *Juniperus* and *Pinus halepensis* can take root.

Seldom visited in the past, except by a few lovers of quiet and of nature, the dune was first of all 'adapted' to receive a pipe line system. This upset the dune profile very severely. Later on a tarred road was built, opening up the dune to thousands of cars and tourists. Because of the effect of levelling out by vehicles parking everywhere at random, of tourists walking over the site, and of wind erosion, the dune began to disappear and with it the sand-dune species.

Before any attempt was made to re-introduce these species, the following problems had to be solved:

1. The encroachment had to be stopped or at least it had to be confined to certain points.

2. Thereafter, the dune biotope had to be reconstituted, more especially by allowing the sand-fixing process to start again.

3. Finally, there was the problem of the multiplication and re-introduction of the threatened species.

Effective and sometimes original solutions were found: the destruction caused by unrestricted access was checked by limiting vehicles to areas levelled out to accommodate them, using rock barriers to prevent random parking, and by fencing off certain areas to keep out walkers—all this to facilitate a rapid restoration of plant-growth. The reconstitution of the dune and the fixing of the sand were brought about by using two native species which, moreover, were on the list of threatened species: *Diotis candidissima* and *Limoniastrum mono-petalum*. Thanks to the planting in large numbers of these two species, the erosion was checked and the stabilizing process restored. The other rare species still growing nearby benefited from this positive development and began to spread.

A number of tests were carried out to determine the conditions best suited to the species it was intended to re-introduce. These species were divided into two groups:

1. Those which were to act as pioneer species, as was the case with *Diotis* and *Limoniastrum*, for which propagation by cuttings proved the most effective and practical method, since collecting of the seeds and direct sowing did not always give good results.

2. Those which would be introduced at a second stage, such as *Matthiola tricuspidata, Matthiola sinuata, Eryngium maritimum* and *Medicago marina*; these were to be multiplied by seeds from intensive cultivation in gardens, or in the case of *Pancratium maritimum*, re-introduced in the form of bulbs.

Sowing seeds of the extremely rare *Myosotis ruscinonensis*, presumed to be extinct in its natural habitat (the beach at Argelès) and which we had begun to multiply in a plant nursery, was also contemplated.

Unfortunately, as if to remind us that the struggle for the protection of nature is never over, men in too much of a hurry took it into their heads to build a dike to protect the nearby road, thereby threatening to undo the work of many years.

DISCUSSION

D. W. JEFFREY (Dublin) asked if there had been any co-operation with the local authorities involved.

L. OLIVIER replied that he had received little direct co-operation from the authorities. Advice given to governments from an international body such as the present conference would help in situations such as the one he had described.

S. WAHLBERG (Sweden) enquired how the speaker's work had been funded.

L. OLIVIER replied that money was provided from his country's National Parks budget.

J. P. M. BRENAN (Kew) praised Olivier's work and enquired about publicity leaflets and posters.

L. OLIVIER felt that although local people (and especially schoolchildren) could be made aware of the significance of the project, it was the tourists who came just for the sand and sun that were to blame and these people were difficult to educate.

PART TWO

National Policies and Activities

Argentina: The Conservation of Endemic and Threatened Plant Species within Botanic Gardens

ELÍAS R. DE LA SOTA

La Plata National University, Argentina

It is the subject-matter of this conference to discuss the rôle of botanic gardens as a means of ensuring the conservation of plant endemics and species in danger of disappearance due to their rarity, to excessive exploitation or to constant alteration of their natural habitats by man. I must confess that I do not work in any of the few botanic gardens that exist in Argentina, but as a naturalist I have a keen interest in everything relating to plants and the conservation of their native habitats.

The purpose of this contribution is to give a brief summary of the number and present state of botanic gardens in Argentina, along with a few suggestions as to what might be done in my country concerning these problems. I have frequently been conscious of Argentina's lack of botanic gardens, and of the important functions that they should fulfil apart from the provision of space for recreation and a purely superficial representation of our native flora.

Even if the university botany departments are endowed with herbaria, libraries and laboratories, there is a constant and increasing necessity for research workers to have at their disposal living laboratories such as botanic gardens, so as to be able to develop biosystematic investigations which cannot be done with dry plant material and microscopes alone.

About sixty years ago, in 1915, Hill wrote: "It is a matter of regret to all botanists that South America, so rich a storehouse of botanical treasures, should contain so few important botanic gardens". This is the sad truth, and even more so in a country as large and as floristically diverse as Argentina; and the situation has changed little since then.

In 1962, Jirásek listed 12 botanic gardens in Argentina. This is far from being exact, as there are only two that can really be considered worthy of that denomination: the 'Carlos Thays' Municipal Botanical Garden of the city of Buenos Aires, and the Castelar Botanical Garden administered by the Ministry of Agriculture.

The first one bears the name of its founder and is located in the very heart of the city of Buenos Aires. It is the oldest botanic garden and has existed since 1898. It is imprisoned by a belt of skyscrapers and avenues with heavy traffic, and occupies an area of a little more than eight hectares. It has a small Botanical Museum and a few greenhouses; the Municipal Gardening School 'Cristóbal M. Hicken' has been there since 1914. The systematic part of the Garden is very small in relation to the total area (which is mostly used as a public park) and is very agreeable and luxuriant though perhaps too densely planted. It is characterized by a rich floristic diversity and exotic trees are preponderant.

The buildings protecting the Garden, and the proximity of the wide La Plata River, have permitted the growth of several trees from the northeastern and northwestern subtropical forests of Argentina. Until very recently, the Municipal Botanical Institute was also there, with a small herbarium (BAJ) and a poor library. But now, these elementary tools for plant determination have been transferred, along with the scant technical personnel, to the Avellaneda Park which is situated in another part of the city of Buenos Aires.

According to a list of Ratera and Montani (1959) there were about 316 native plant species curated in the Garden. At present Antonio M. García, Head of the Municipal Botanical Institute, is preparing a catalogue which he kindly allowed me to consult. A superficial analysis shows that the native flora is poorly represented; for example, the great majority of the 69 species of Aizoaceae listed are of South African origin and only 2 of the 100 conifer species are native. For the visitor to the Garden, it is very difficult or practically impossible to get even the slightest idea about the diversity of the indigenous flora and the most significant vegetation types of Argentina.

The second botanic garden mentioned, that of Castelar, belongs to I.N.T.A. (National Institute of Agricultural Technology) and is located near the city of Castelar, which is close to Buenos Aires. It covers 20 hectares, half of which contains plant species arranged in systematic order. The other half contains examples of the most characteristic phytogeographic regions and ecological habitats (salt, sand, rock, marsh communities, etc.). This Garden was started in 1947, but it came to a standstill and further progress seems to be hopeless at the present.

The Castelar Garden is for acclimatization and has no function for recreation or conservation. It co-operates with the University, especially with the Schools of Agronomy, in providing space and botanical material for research projects and teaching. Taking into account the structure and aims of the institution that created the Garden, these interests are eminently practical.

In the only published catalogue (Milano, Rial Alberti and García, 1962) listing the plants grown, one can see the rich collections of *Pinus, Juniperus, Salix, Populus* and *Eucalyptus*. According to information given to me by the technical experts responsible for the Garden, 4000 species both woody and herbaceous are maintained, of which only approximately 10 per cent are indigenous. While it is possible to observe healthy stands of *Araucaria angustifolia* (the well-known Parana Pine intensively exploited for the timber industry), in contrast *Araucaria araucana*, the 'Pehuén', which is found in a reduced sector of the Patagonian-Andean forests, grows very poorly.

Though the Castelar Garden does not present the space limitations of the 'Carlos Thays' Garden, its climatic and edaphic conditions are far from being the most suitable. The action of the La Plata estuary on regulating minimum temperatures does not affect it, and moreover the presence of a layer of 'caliche' close to the surface of the ground harms the growth of several tree species.

Both botanic gardens exchange seeds with similar institutions in different parts of the world.

Finally, in several Argentine universities there are very reduced gardens which should more properly be called 'systematic gardens', functioning exclusively for teaching purposes; these include the Faculty of Agronomy of Buenos Aires, the Faculty of Agronomy of La Plata, and the 'Chacras de Coria' Garden attached to the National University of Cuyo (in Mendoza), which includes 423 species in cultivation according to a list published by Ambrosetti (1965). The 'L. R. Parodi' Garden in Santa Fe, created in 1971, should also be mentioned.

Argentina is a big country, ranging from 22°S. to 55°S. and with elevations from sea-level to 7000 m. There is therefore a vast range of natural parameters, some of which (such as temperature and daylight) cannot be controlled or artificially reproduced in botanic gardens except under very special circumstances.

As we have seen, the only true botanic gardens that exist in our country are located in or very near the city of Buenos Aires, on the plains far from the areas which attract most attention from the conservation viewpoint. These few gardens have not had much success with the cultivation and curation of acutely threatened species. In 1965 the Organization of American States (O.A.S.) published a list—very incomplete in our judgement—of these species, among which were *Alnus jorullensis, Podocarpus andinus, Podocarpus nubigenus, Araucaria araucana, Drimys winteri, Nothofagus obliqua, Saxe-Gothaea conspicua, Euterpe edulis* and *Persea lingue*. Even the majority of these are not represented in either of the two botanic gardens mentioned.

The botanic gardens must fulfil a diversity of objectives, besides the traditional recreative, ornamental or educational ones. In these latter senses they are similar to public parks, though more controlled, more diversified floristically and with their plants correctly identified. But in a country like Argentina with enormous climatic and phytogeographical diversity, the main problem is where to site the botanic gardens of the future in order to combine the conservation of the native regional flora with the proper use of facilities, so that they may fulfil the many functions which the two gardens mentioned above are not able to perform.

Turning now to the problems discussed at this conference, we all know of the importance in conservation of National Parks and nature reserves. It is evident that botanic gardens can fulfil a fundamental rôle in the conservation of plant species, facilitating at the same time the development of experimental studies. This kind of botanic garden would usually be much smaller than the classic ones but would have the adequate control, and the possibility of maintaining, studying and multiplying important floristic elements, not only of endemics, but also of rare species or those on the verge of extinction.

This statement implies the necessity of the creation of a sufficient number of conservation-orientated botanic gardens. In general their location, dependence, organization and scope should fulfil the following conditions:

1. They should be located near botanical research centres, preferably attached
    to the universities and, if possible, in hilly regions where there is a greater
    range of habitat because of altitude, aspect and situation (water, marsh,
    woodland, etc.).
2. They should be organized from the ecological and phytogeographical view-
    point, with a regional emphasis, and relegating ornamental and landscape
    aspects to a subordinate place.
3. They could also contain biological stations to work in co-operation with
    university research and teaching centres.
4. The administration, control and utilization should be related to these research
    centres, and the gardens should not be allowed to become agricultural
    experimental stations or commercial nurseries.
5. They should conserve endemic, uncommon or threatened plants, indicate
    priorities for conservation, and cultivate the characteristic species of the
    region. These could be obtained from the neighbouring National Parks and
    nature reserves, and by botanists working on regional floras who are able to
    collect living material.
6. Such species should be catalogued.
7. Surplus plant material should be distributed to accredited, national or foreign
    botanic gardens, public parks, horticultural societies, etc.
8. They should initiate and stimulate the development of basic research.
9. They should provide material for teaching purposes.

The function indicated in (7) would help to publicize those species that require
greater protection, being a simple and effective way of conserving them and
stimulating the cultivation of the national flora.

As already suggested by Mickel (1977), besides distributing the critical species
between botanical institutions, a further positive factor in conservation is to
introduce them into general horticulture. Mickel gives some examples of Pteridophyta, a group on which he has been working. For example, *Didymochlaena
truncatula* is now found more frequently under cultivation than it is in its
natural habitat. *Ginkgo biloba* is another outstanding example which needs no
further comment. Numerous species which are rare or even under threat of
extinction that have some ornamental attraction could be thus popularized, at
least regionally.

The conservation aspect of botanic gardens could act as a very effective link
between the National Parks and the constantly reduced but still undevastated
areas on the one hand and the people on the other. If there was a way whereby
botanic gardens could provide plant material so preventing uncontrolled losses
from the wild, which are often caused by individuals trying unsuccessfully to
grow plants which have not been acclimatized properly due to inadequate
knowledge, they would play an essential intermediate rôle.

The necessary regionalism of these gardens would undoubtedly make their
rôle successful and economically possible. The controlled popularization of many
endangered species could thus be turned into one of the simplest instruments to
achieve the objectives of conservation.

I wish to express my gratitude to Dr Ovidio Núñez for his constant interest and very valuable assistance during the preparation of this brief contribution and also my thanks to Dr Genevieve Dawson for the English version of the original manuscript. I would also like to thank the organizers of this Conservation Conference of the Royal Botanic Gardens at Kew, British Caledonian Airways and the Argentine National Research Council (C.O.N.I.C.E.T.), for making possible my attendance at this important meeting.

REFERENCES

AMBROSETTI, J. A. (1965). Catálogo de las plantas fanerogámicas del Jardín Botánico. *Bol. Jard. Bot. Chacras de Coria* **1**: 1–31.

CUGNAC, A. DE (1953). Le rôle des jardines botaniques pour la conservation des espèces menacées de disparition ou d'altération. *Ann. Biol.* **29**: 361–367.

DE FINA, A. L. (1942). Las sierras de Mar del Plata, región apropiada para instalar el Arboretum Nacional Argentino. *Rev. Argent. Agron.* **7**: 188–192.

HILL, A. W. (1915). The History and Functions of Botanic Gardens. *Ann. Missouri Bot. Gard.* **2(1–2)**: 185–240.

JIRÁSEK, V. (1962). List of Addresses of Botanic Gardens. *Acta Horti Bot. Pragensis*, **1962**: 11–66.

MICKEL, J. T. (1977). Rare or endangered Pteridophytes in the New World and Their Prospects for the Future. *In 'Extinction Is Forever'* (eds G. T. Prance & T. S. Elias). New York Botanical Garden. Pp. 323–328.

MILANO, V. A., F. RIAL ALBERTI & A. L. GARCÍA (1962). Catálogo de las especies cultivadas en la sección sistemática del Jardín de Aclimatación de Castelar. *Instituto Nacional Tecnología Agropecuaria (I.N.T.A.)*, Misc. No. **46**: 1–81.

ORGANIZACIÓN ESTADOS AMERICANOS (O.E.A.) (1965). Lista de especies de fauna y flora en vías de extinción en los estados miembros Argentina, Bolivia y Ecuador. *Conferencia Especializada Latinoamericana Problemas Conservación Recursos Naturales Renovables Continente*, Doc. **16**: 1–8.

RATERA, E. L. & R. G. MONTANI (1959). *Plantas de la flora argentina cultivadas en el Jardín Botánico 'Carlos Thays'*. Municipalidad Ciudad Buenos Aires. 22 pp.

THAYS, C. L. (hijo) (1929). El Jardín Botánico Municipal de la Ciudad de Buenos Aires. *Intendencia Municipal Ciudad Buenos Aires*. 129 pp.

WENT, F. W. (1953). Remarks about the relationship between botanical gardens and scientific research. *Ann. Biol.* **29**: 455–456.

Australia: The Cultivation of Native Endangered Plants in Canberra Botanic Gardens

J. W. WRIGLEY

National Botanic Gardens, Canberra, Australia

Canberra Botanic Gardens is a unique institution dealing solely with the Australian flora. It occupies 45 hectares of the slopes of Black Mountain, a eucalypt-clothed hill reserve 2 km from the centre of Canberra. The eucalypts provide a background for the informal layout of the Gardens in which about 5,000 species of Australian native plants are in cultivation. The policy is to maintain an even balance between the three main functions of the Gardens—Recreation, Research and Education. Activities in these three main areas will be outlined below.

Canberra's climate is harsh by Australian standards with severe winters (minimum temperatures may reach −10°C.) and hot dry summers (maximum 39°C.). The rainfall is 665 mm distributed reasonably evenly throughout the year. A coastal annexe at Jervis Bay with a milder climate has been established to enable a wider variety of plants to be grown outside. Even with this facility, important tropical collections of orchids, ferns and other rain forest plants are maintained in heated glasshouses.

Material grown in the Gardens is collected from the field throughout Australia by horticulturists and botanists. A reference herbarium housing some 70,000 specimens ensures that all plants collected from the field are correctly determined.

Plants in the open Gardens are arranged in four different categories:

1. Taxonomic groupings, in which plants from the same family or even genus are grouped together.
2. Ecological groupings in which plants occurring in similar environments are grouped, e.g. rain forest.
3. Educational groupings, in which species may be grouped because of a common function, e.g. plants used by Aborigines, native plants considered weeds in other countries.
4. Aesthetic groupings, in which plants are used for their value to the landscape, e.g. around buildings, carparks, etc.

With over 300,000 visitors per year, the visiting rate is very high when it is considered that Canberra's population is a mere 220,000. A recent visitor survey showed that approximately 11 per cent of visitors are from overseas and that about 50 per cent are from other Australian states. The Gardens are still young although the concept for the Australian Botanic Gardens dates from the 1913 Walter Burley Griffin plan for Canberra. The first tree was planted in 1949 and further development was very slow until the late sixties, when the Gardens were fully appreciated as a national asset and afforded a separate working budget. Development accelerated for the official opening in 1970 by the Prime Minister, the Rt. Hon. J. G. Gorton, as part of the proceedings of the 6th International Congress of Park Administration. Since this time, further expansion has been steady with extended field trips to North Queensland, the Kimberley Ranges (Western Australia), the Grampian Ranges (Victoria) and several others. Budget restrictions over the past two or three years as a result of Government economies have slowed down plans for a large tropical conservatory, further service glasshouse space and expansion into an adjacent 30 hectare area.

RECREATION

The Gardens are open from 9 a.m. to 5 p.m. every day of the year except Christmas Day and afford a pleasant, quiet noise-free atmosphere in which to relax or take a quiet stroll. In spring and early summer, when flowering is at a peak, the Gardens are alive with colour. At other times when flowers are less prolific the soft greens and delicate perfumes of the Australian 'bush' can be enjoyed. The native birds, including colourful parrots and honeyeaters, kookaburras and small wrens, are always a delight to those prepared to be quiet and enjoy their activities at feeding and nesting times. A special leaflet is produced to help visitors know these birds a little better.

Recreation is essentially passive, as there is no space for ball games or bicycle riding. These are well provided for elsewhere in the city. Rolling lawns are few although there is adequate provision for larger school parties to have lunch in attractive surroundings with expansive views over the city. All plants are clearly labelled, and with native plants riding on a crest of a wave of popularity, home gardeners use the Gardens extensively for ideas of new species to grow. Weddings are frequently held in the Gardens particularly in the contrived rain forest area where birds are always active and the temperature is mild in all seasons.

The rain forest, an ecological grouping, deserves special mention. It has been established in a deep dry gully sculptured from the slopes of Black Mountain. High humidity has been created by the provision of hundreds of misting nozzles controlled by a time clock. These are activated for two minutes every four minutes (less frequently in cooler weather) and produce a fine mist which fogs the gully and provides an optimum atmosphere for plant growth. Because of its topography this area is virtually frost-free. Many temperate rain forest species are now well established and are forming the canopy so essential for the welfare of understorey plants.

Two general routes have been signposted through the Gardens. The first, known as the White Arrow Walk and designated by a series of white arrows, leads the visitor through some of the older, well established sections of the Gardens. It is about 1.5 km of generally easy walking although some steep steps are encountered in the rain forest. For those with strollers, or for those unable to negotiate the steps, this section may be bypassed and viewed from several vantage points.

The natural white-barked *Eucalyptus mannifera* subsp. *maculosa* form a most attractive backdrop to many parts of this walk. The old open terraced tracks have recently been redesigned to provide a more intimate walk, with quiet seating areas along the way.

The major families seen on the track are:

1. Myrtaceae, well represented by *Leptospermum* spp., *Melaleuca* spp., and *Callistemon* spp., which are best seen in November and early December.
2. Proteaceae, with a large collection of *Grevillea* spp., some *Banksia* spp. and a considerable representation of *Hakea* spp. Some of these may be seen in flower at most times of the year.
3. Mimosaceae, an excellent collection of *Acacia* spp., which starts to flower in late winter and continues into spring.

Several side tracks or detours can be made to see areas of Asteraceae and Lamiaceae. Both of these sections are in full flower in mid-October and the *Prostanthera* spp. in the latter section are always particularly spectacular at this time.

The other walk is known as the Blue Arrow Walk. This track branches off about half-way around the White Arrow Walk and enters areas of newer plantings. An additional 1.5 km is added by this walk but again the grades are generally easy. There is much to interest the enthusiast in this area as many rarer and more frost-tender species have been established. An important collection of *Boronia* spp. may be seen flowering in mid-October. The rare *Boronia subulifolia*, *B. serrulata* and the cultivar *B. mollis* 'Lorne Pride' are outstanding. Other genera within Rutaceae such as *Phebalium*, *Asterolasia* and *Correa* may also be seen, the last mentioned providing winter interest. A second *Acacia* section is becoming well established and here the rare *A. denticulosa* may be seen. Other sections of Proteaceae, Solanaceae, Sterculiaceae and several smaller families add interest to this tree-covered walk. To do the White Arrow and Blue Arrow Walks comfortably, the visitor should allow half a day. Ample seating is provided and a cup of tea can be purchased at the snack van on emerging from the rain forest.

RESEARCH

A well appointed horticultural research laboratory was built in 1970 and although hampered by staff restrictions some outstanding work has been published since that time. Research work has been essentially of an applied nature solving problems directly associated with the cultivation or propagation of native plants.

Major projects have been concerned with the following areas:

1. Germination and cultivation of Australian terrestrial orchids.
2. Tissue culture of *Anigozanthos* (Kangaroo paws) and *Macropidia* as well as several other genera.
3. Breaking of seed dormancy in Rutaceae and Proteaceae.

4. Nutrient requirements of Western Australian *Banksia* spp.

5. Grafting of native plants for disease resistance.

6. Propagation of ferns using aseptic techniques.

7. Use of growth promoting substances in the rooting of difficult cuttings.

Taxonomic research has been limited but considerable use has been made of the herbarium by visiting botanists for research purposes.

The nursery is a semi-research unit, as here plant material is brought in from the field and propagated from either seed or cuttings. Commonly, species are being propagated domestically for the first time. Decisions have to be made as to what conditions may lead to the best success. Consideration has to be given to the habitat from which the species was collected and thus the recording of field data is an important factor in the total propagation exercise.

Specially designed seed and cutting rooms have been constructed. Soil sterilization is achieved by passing a steam-air mixture through potting soils to kill pathogens. Particular attention has been given to the elimination of the root-rot fungus *Phytophthora cinnamomi* from the nursery, and soil treatment together with other hygiene techniques have virtually achieved this. Three service glasshouses, together with borrowed glasshouse space at the Government's Yarralumla Nursery, house Australia's best native orchid collection, a good fern collection and many other tropical plants collected from Northern Australia. The orchid collection deserves special mention as the epiphytic representation is almost complete and the terrestrial collection is being built up at a steady rate. This latter group is rarely seen in cultivation and includes some unique flowers.

EDUCATION

Education should be a major function of any botanic garden and at Canberra it is given the prominence it deserves.

Labelling of Plants. All plants are labelled with engraved, anodised aluminium labels. These are permanent and are virtually vandal-proof. Larger stained wooden signs are displayed in family beds explaining in lay terms points of interest concerning the family.

Display Room. A display room illustrates various aspects of the Australian flora, with full use being made of live material and three-dimensional models. These displays are changed quarterly and cover such subjects as seed dispersal, economic uses of Australian plants, Aboriginal uses, vegetation types, aquatic plants and many more.

Publications. An annual series entitled *Growing Native Plants* is produced expressly for the home gardener and landscaper. This has proved very popular and volume eight is currently in press. Several other publications are also available dealing with aspects of the Gardens' activities.

Guided Tours and Specialist Groups. Trained rangers as well as technical and professional staff take school and adult groups for various instruction. Some 7,000 schoolchildren were shown various facets of the Gardens last year. Some were given general tours but teachers are encouraged to concentrate on narrow fields that fit into school curricula. Examples of special interest groups are rain forest ecology and the family Myrtaceae for senior high school students, nursery techniques for university forestry students and nesting habits of birds for younger children.

Children's Study Groups. Senior primary students are instructed in groups of eight to twelve in techniques of basic horticulture using native plants. These classes proceed for a school term and usually consist of about ten weekly sessions of one and a half hours. Children sow seed and prepare cuttings and on the last lesson take home the plants produced.

Nature Trail. A nature trail has been developed in an unwatered natural forest section of the Gardens. Clear descriptive photo-metal signs are used to explain natural phenomena and are used in conjunction with an illustrated leaflet. This trail explains the inter-relationship of organisms in a simple and interesting fashion and children of all ages as well as adults enjoy this somewhat different approach to this fascinating subject.

Aboriginal Trail. Another trail through the developed sections leads the visitor past plants used by the Aborigines. A well illustrated leaflet explaining those uses adds to the value of this walk.

Canberra Botanic Gardens houses the largest collection of Australian plants anywhere in the world. It is in fact a living museum set in a scenic, quiet, informal atmosphere where the unique beauty of the Australian flora can be appreciated and studied.

(Canberra Botanic Gardens was renamed National Botanic Gardens on 21 December 1978.)

Czechoslovakia: The Rôle of Botanic Gardens in the Conservation of Rare and Threatened Plants in Slovakia

LADISLAV ŠOMŠÁK

Botanical Garden, Faculty of Sciences, Comenian University, Bratislava, Czechoslovakia

Changes in the original, natural vegetation of all the countries of Europe as a result of technical advances have become very apparent in recent decades. The Slovak Socialist Republic is no exception and these changes have been very significant during the last twenty years. Industrial development with its accompanying phenomena has caused not only a decrease in the area of natural stands of vegetation, but at the same time has diminished the distribution of numerous individual plant species. In the early fifties some plants were already so rare that they were declared protected plants by law. The regions with many plants or those of high botanical importance were declared as natural reservations. By 1971 more than 110 botanical reserves and protected sites had been declared in Slovakia (Mihálik *et al.*, 1971).

The major work in protecting the Slovak flora is carried out by the Slovak Institute for Nature Conservation in Bratislava. During the years 1963–1972 all regions in Slovakia were mapped from the conservation viewpoint and therefore much information was obtained on the needs of the protected and rare plants. The most important floristic riches of our country are protected in the National Parks of the High Tatra, the Pieniny and the Low Tatra as well as in the protected landscape regions of which there are seven in Slovakia. They cover the most valuable regions from the plains up to the subalpine zone. The protective laws concern only a relatively small group of plants—about 110 taxa, mainly from those regions where the pressures of industrial development are not so evident and the danger appears mainly from the effects of tourism.

The most threatened plants in our country today are those taxa which were still widespread twenty years ago and which therefore escaped the attention both of botanists, and of the authorities of the State Nature Conservancy. The species mainly affected in this way covered the plains and hill regions. The changes to their habitat and natural environment came about under the influence and the development of agriculture and industry. These changes we can characterize briefly as follows:

1. The regulation of the rivers flowing through the largest Slovakian plains of the Danube, Tisza and Záhorská. This regulation caused changes in the water regime, mainly removing the meanders, accelerating the discharge of water and decreasing the ground water level.

2. The drainage of the wetter soils through a grid of surface canals and underground drains; the latter have also affected the wet sloping meadows of the hill country and submontane regions.

3. The conversion of the drained meadows to arable land.

4. The destruction of the shrub communities in the agricultural landscape as a consequence of tilling soils by mechanical farming methods.

5. The introduction of large quantities of industrial fertilizers with the aim of increasing agricultural production, also the application of herbicides and pesticides.

6. The reduction of natural stands of lowland vegetation and flora due to urbanization and the construction of industrial centres and their expansion, which at present mainly occurs in the lowlands.

7. Similar effects have also been caused by the construction of the main communication grids and the building of vast reservoirs and hydro-electric power-stations.

8. A further considerable change in the structure of both the lowland and upland landscape has been caused by the explosion of leisure activities, in particular by the construction of individual cottages. This is affecting considerable areas of forest edge, river banks and other sites that were previously safe from agricultural development.

9. Forest management has had an influence on the localities of many plants. These sites are mainly the small enclaves of natural vegetation on the secondary steppes of clay, sand and rock. Their local and scattered nature and small sizes were negligible from an agricultural viewpoint and therefore in the majority of cases these areas were forested. In several regions of Slovakia, mainly on sandy soils, forest plantations have been created over large areas. The preparation of the soil for planting involves heavy machinery passing over the land and this essentially changes the condition of the soil.

10. The changes in the original quality of the soil nutrients in the plains and uplands by pollution from the large industrial centres have in several cases practically destroyed the soils leaving a minimal vegetation cover.

Phytosociological studies of the vegetation of the alluvial meadows were among the first to show losses in the flora. These studies were carried out by the Department of Geobotany, Faculty of Sciences, Comenian University, Bratislava, since 1963. An early solution was to establish nature reserves. This was a great step forward in the protection of the plant riches in the affected regions of Slovakia; however it was impossible to protect all threatened plants in this way. The grid of nature reserves or protected areas, however planned, would be a serious obstacle to the rapid development of agriculture and industry. Similarly the suggestion for the widespread listing of protected plants by law has had little effect.

Slovak botanists welcomed the idea of growing and cultivating threatened and rare plants in botanic gardens, in support of the resolutions of the plenary session of I.A.B.G. in Moscow (1975), the International Botanical Congress in Leningrad (1975) and in accordance with the European project of the IUCN

Threatened Plants Committee at Kew (Lucas & Walters, 1976). In 1975 we started detailed research on the threatened and rare plants of the Slovakian flora concentrating on the plains and uplands as these are the most disturbed regions. The aims of this research was to discover the degree of threat to species and the possibilities for cultivating these plants in the botanic gardens of the Comenian University. The whole problem was discussed by the Slovak Botanical Society in 1977, and the topic was made the main theme at a seminar under the title 'Natural Sciences in the Environment', held by the Faculty of Sciences, Comenian University, in April 1977.*

The number of threatened and rare species from the plains and hill country was surprisingly high, as can be seen from Table 1.

TABLE 1

Analysis of Threatened and Rare Plants of Slovakia

	No. of taxa
1. Plants of slow flowing and stagnant water	20–25
2. Plants of swamp and bog biotopes	55–60
3. Plants of the plains and hill country meadows	30–35
4. Sand plants	40–45
5. Halophytic plants	25–30
6. Plants of secondary clay and rocky steppes	90–95
7. Plants of fields and ruderal sites	40–45
Total	300–335

As indicated on the table, the largest number of threatened and rare plants comes from the secondary clay and rocky steppes. The survival of these plants is threatened by forest planting which is often carried out with unsuitable trees, by the construction of individual cottages, and by the establishment of vineyards and small gardens. The number is high in this group because it also contains those species which are not threatened by the influence of man, but are rare in nature anyway. For these plants cultivation in the botanic garden will provide a certain insurance for the future. Such species include *Campanula macrostachya*, *Crepis pannonica*, *Doronicum hungaricum*, *Echinops ritro*, *Echium russicum*, *Muscari botryoides*, *Ophrys holosericea* (= *fuciflora*) and *Pedicularis comosa*. These species pose a problem in the determination of their natural occurrence (Dostál, 1950).

There is also a relatively large number in the group of threatened bog plants. Their biotopes are threatened mostly by new drainage schemes. The change of water-soil balances has caused the retreat of several plants and the majority are in addition on the edge of their range; examples are *Gladiolus palustris*, *Juncus atratus*, *Liparis loeselii*, *Senecio congestus* and *Typha minima*.

The same is true if not quite so critical with some meadow plants. For example in our country the once very common species *Achillea ptarmica* is today

* The proceedings of this conference will be published in the near future.

only positively identified from one locality with only a few individuals. The same problem exists with some species from the alluvial meadows, for example *Gentiana pneumonanthe*, *Serratula lycopifolia* and *Dianthus superbus* subsp. *superbus*.

The psammophytic Slovak flora has begun to suffer more recently. The intensive utilization of the soil has pushed back many of the species from large open sandy sites to forest borders. Some species are very reduced and now confined to small open sites in forests; examples are *Colchicum arenarium*, *Iris arenaria*, *Onosma arenaria*, *Silene conica*, *Peucedanum arenarium*, *Tribulus terrestris*, *Trigonella monspeliaca*. Many halophytic plants in the Slovak Socialist Republic are located in nature reserves, and from this point of view are not threatened to the same extent by loss of habitat. However experience indicates that outside influences close to the reserves can seriously disturb the vitality of some species, e.g. *Iris spuria*, *Limonium gmelinii*, *Plantago tenuiflora*, *Ranunculus pedatus*, *Beckmannia eruciformis*. The already small areas of halophytic vegetation that exist at present and the possibility of potential threats to their long-term survival was the main reason for including the majority of halophytes of Slovakia in the list of threatened and rare plants. After the experience of the last 10 years, some field and ruderal species were added to the list. Their decline was brought about by the industrialization of agriculture.

Besides the relatively high number of taxa from every habitat covered in the Slovakian threatened and rare plant list, it is necessary to mention that there are species which are not threatened throughout their whole area in Slovakia, but only in some particular regions. These plants have also been included on the threatened list in order to protect their differing populations from various environmental threats. The list of threatened plants from Slovakia (Šomšák, 1977) therefore included threatened regional populations as well.

The second phase of research deals with threatened and rare plants of Slovakia in the submontane-montane and subalpine-alpine zones. After a preliminary evaluation of the situation, Šomšák *et al.* (in press) came to the conclusion that more than 100 taxa should be covered. The majority of them are very rare; examples are *Cystopteris regia*, *Eriophorum gracile*, *Gentiana nivalis*, *Ligularia glauca* and *Trichophorum (Scirpus) pumilum*. Other species may disappear under the influence of natural succession, or by the afforestation of small meadow enclaves in the mountain and subalpine zone. The most serious threat to these species is from various recreational activities, principally tourism; trampling is a part of this problem, mainly in the High Tatra region where pressure from the increasing number of tourists is reaching the carrying capacity of the land. The results of the first observations dealing with changes in this region are truly alarming (Šomšák *et al.*, in press).

Preliminary results from this whole study show that from the Slovak Socialist Republic about 400 plant taxa are threatened or rare and these should receive priority attention to preserve them as part of our scientific, historic and cultural heritage. As mentioned above the protection of these plants by law in the majority of cases is impossible if for no other reason than from the changes that have and are occurring to the total environmental conditions in which they live. Therefore the cultivation and conservation of these plants in botanic gardens must be a task of high priority in Slovak botany.

What are the possibilities for this work? The main task of conservation of the genetic stock belongs to the classical botanic gardens, such as that of the Comenian University with its two departments, the first of which is situated in Bratislava, the capital of Slovakia. During its 40 years, this garden has dealt with the cultivation of wild source material from the West Carpathians, wherever suitable for our climate. Although the protected and threatened species are of particular importance, this garden is not suitable for growing many of them due to the considerable pollution from the city. Accordingly a new botanic garden was established in 1967—this is the second department. It is situated approximately 20 km north of Bratislava at the village of Stupava at the bottom of the Malé Karpaty mountains, between 200 and 250 m above sea-level. The total area of the garden is 24 hectares. It is situated in the forest where climatic conditions are good and there is a diversity of soils suitable for cultivation of plants from the plains and hill regions. It appears that it will be possible to grow practically all the threatened and rare species here. We will start with the cultivation of these species from the Slovak plains in 1979–1980 on an area of about 5 hectares.

Introductory projects have shown up many problems, for example the low germination rate of seeds of some species and the establishment of conditions suitable for some specific groups, such as halophytes. Positive plans for planting populations in particular ways such as in flower-beds, free plantings, or in biological or aesthetic groupings will be drawn up in the best traditions of landscape and garden design to reach the main goal that this garden will become the cultivation and conservation centre for threatened and rare plants of the plains and hill regions.

The threatened and rare species of the submontane and montane zone (600–1400 m) will be concentrated in another botanic garden of the Comenian University; this is situated in the centre of Slovakia, at Turčianska Štiavnička near Martin, approximately 240 km from Bratislava. With its climate and location in the centre of the Fatra mountains it is very suitable for growing these species. There are definite plans to grow about 70 populations of the appropriate taxa.

The rare and threatened plants of the subalpine and alpine zone (1600–2656 m) will be concentrated in a specific botanic garden which belongs to the High Tatra State National Park. This garden has been established in the High Tatra at Štrbské Pleso (1365 m), and works in close good co-operation with the Comenian University.

The cultivation and conservation of the threatened and rare plants of Slovakia in the botanic gardens is a most important task. The main goal is to obtain a detailed knowledge of the autecology of particular plants, enriching our understanding of the spread of their populations, their vitality in the original sites, their place in phytocoenoses, seed germination, as well as in the possibilities of cultivating them for aesthetic and other purposes.

At the same time, however, conservation is an important theme for the education of the young botanical generation. Already this year we have started propagating this theme with the public. Published in 1978, for all levels in schools, is a Flora with colour pictures covering mainly the threatened and rare species and supplemented by a text in part explaining their historic, scientific and cultural importance. The first chapter deals with the protection of plants of Slovakia (Šomšák & Slivka, 1978).

The directorate of the botanic gardens of the Comenian University thanks the initiators of this conference and the IUCN Threatened Plants Committee at the Royal Botanic Gardens, Kew. Finally, allow me to assure you that every effort will be made to realize this very important task of plant conservation.

REFERENCES

DOSTÁL, J. (1950). *Květena ČSR*. Přírodovědecké nakladatalství, Prague.

LUCAS, G. LL. & S. M. WALTERS (1976). *List of Rare, Threatened and Endemic Plants for the Countries of Europe*. Mimeo. (Subsequently published by the Council of Europe as *'List of rare, threatened and endemic plants in Europe'* by IUCN Threatened Plants Committee. Nature and Environment Series No. 14. Strasbourg, 1977.)

MIHÁLIK Š. *et al.* (1971). *Chránené územia a prírodné výtvory Slovenska*. Vydavatelstvo Príroda, Bratislava.

ŠOMŠÁK, L. (1977). *Ohrozené a zriedkavé taxóny horských a vysokohorských polôh Slovenska*. Bratislava. (The threatened and rare taxa of the mountain range of Slovakia).

ŠOMŠÁK, L., F. KUBÍČEK, I. HÁBEROVÁ & E. MAJZLÁNOVÁ (in press). *Influence of tourism on the vegetation of High Tatras*. Biologia VSAV, Bratislava.

ŠOMŠÁK, L. & Š. SLIVKA (1978). *Chránené rastliny Slovenska*. ČSTK-Pressfoto, Bratislava. (The protected plants of Slovakia).

TAKHTAJAN, A. L. (ed.) (1975). *Red Book: Native Plant Species to be Protected in the USSR*. Leningrad. (In Russian).

India: Botanic Gardens and Threatened Plants—A Report

S. K. JAIN

Botanical Survey of India, Howrah, West Bengal

Till the middle of this century, India had one large internationally known botanic garden, namely the Indian (formerly Royal) Botanic Garden near Calcutta, and a few smaller botanic gardens, such as the Lalbagh Botanic Garden at Bangalore, Empress (Victoria) Botanic Garden at Pune (formerly Poona), and the Lloyd Botanic Garden at Darjeeling. The latter three gardens almost confined their activities to ornamental plants and only a small number of exotic trees and other plants were grown. Some good gardens of the past, such as the Botanic Garden at Saharanpur in northern India, had also declined by this time.

In 1953 the government developed another large National Botanic Garden around the nucleus of a historical Nawab's Garden called Sikanderbagh at Lucknow. With the re-organization of the department of the Botanical Survey of India in 1956, plans were drawn up for the development of several botanic gardens. The Calcutta garden was transferred to the management of the central national government, and new experimental botanic gardens were established at Shillong in the eastern Himalayas, Pauri in the western Himalayas, near Pune on the Deccan Plateau, at Yercaud in the Shevaroy hills in south India and at Allahabad in the Gangetic Plains.

The programmes of all these botanic gardens now include the study and conservation of rare, threatened and endemic plants and the establishment of germ plasm banks of certain groups.

Because the flora of all parts of the country has not been fully explored and studied, precise data on rare and threatened plants is still wanting. About 15 years ago, a tentative list of about 150 such plants was drawn up. We realize that many more species are to be added to the list, and it is not improbable that as many as 1000 species may be endangered or vulnerable today. Critical studies on distribution and taxonomy have been done only in a few small families and genera.

In the principal and largest botanic garden of the country, the Indian Botanic Garden near Calcutta, many threatened, rare or endemic species have been brought into cultivation. These include *Nepenthes khasiana, Blachia andamanica, Wallichia densiflora, Bentinckia nicobarica, Ephedra foliata, Platycerium* and a number of orchids.

The Garden at Shillong puts particular emphasis on the collection, study and conservation of orchids. About 1200 species of orchids are reported from India,

113

and about 600 come from eastern India alone; at the Shillong orchidarium, we have been able to collect and grow about 400 species. For some species, such as in the genera *Coelogyne* and *Paphiopedilum*, we have developed very large populations. *Paphiopedilum fairieanum* has been cultivated simulating its natural habitat; it is grown with its associate *Bergenia ligulata* under pine trees in the grounds and also separately in pots.

A number of rare and endangered species are growing in the Shillong garden; the following are noteworthy: *Campanula khasiana*—an endemic species; *Nepenthes khasiana*—an endemic carnivorous plant; *Thunbergia coccinea*—an interesting ornamental climber; *Platycerium* (Staghorn Fern).

The national orchidarium and experimental garden under the southern Circle of the Botanical Survey of India at Coimbatore have a number of rare and endangered species. The following can be mentioned as examples: *Acanthephippium bicolor, Anoectochilus elatior, Pleïone praecox, Paphiopedilum insigne, Sirhookera lanceolata, Psilotum nudum, Lycopodium macrostachys, Osmunda regalis, Ensete superbum, Aristolochia labiosa, Lilium neilgherrense, Vernonia shevaroyensis* and *Bentinckia condapanna*.

A close study of the genus *Coelogyne* showed that three species are perhaps extinct now. These are *Coelogyne albolutea* from mountains of north India, with the only type in the Kew Herbarium, *C. treutleri* from Sikkim, also with the only type being in the Kew Herbarium, and *C. assamica* from Assam.

The flora of India, particularly in the Himalayas and in the Andaman Islands, has a very high level of endemism. We estimate that about 4000 to 5000 plants, out of an estimated 15,000 species in the Indian flora, may be endemic. With such large numbers, no one garden in the country can grow them all for reasons of space and climatic conditions. The regional gardens of the Botanical Survey and the National Botanic Gardens at Lucknow have a programme for maintaining live material of as many endemic and rare species as possible.

The flora of the Andaman and Nicobar islands needs a special mention. It has affinities with the Malaysian flora, and many species found in the Andamans and Malaysia do not extend to the mainland of India. To conserve the endemic and other rare flora, the government are starting a botanic garden near Port Blair; this will be the first botanic garden in the islands. Initially it will be small, but we hope to acquire some virgin forest of several hundred hectares in an area adjacent to the garden.

It is relevant to narrate briefly the present activities and overall programmes of conservation in India. Fortunately almost all our conservation programmes have received support from central and state governments. Except for stray instances, no large scale resistance has so far come forth against any conservation programme.

In many areas of work related to the conservation of flora, we are at the stage of study and evaluation; but for certain groups or species such as the whole of the Orchidaceae and about another 100 species of phanerogams, positive action has been taken by way of a ban or restriction on their collection, sale or export, and by establishing gene sanctuaries.

There is an Indian Board for Wild Life (IBWL) in the Ministry of Agriculture. Originally the Board concerned itself only with animals, but a Flora Wing under the Chairmanship of the Director of the Botanical Survey of India has now been functioning for several years. The Government have a National Committee on Environmental Planning and Co-ordination (NCEPC). It is headed by a very senior and distinguished botanist of the country, Professor B. P. Pal, F.R.S.

The Man and the Biosphere programme (MAB) is also operating in India. It acts as an advisory body and has some funds at its disposal for supporting relevant and appropriate research. Its Committee is now working on the selection of suitable sites for the establishment of Biosphere Reserves. Proposals for selecting forest areas in the country were invited and in the first preliminary assessment, about 50 areas from different vegetation types and ecosystems have been suggested. A small task force is now scrutinizing these proposals.

There are 137 wildlife sanctuaries and only about a dozen recognized botanical gardens in the country. Nine wildlife sanctuaries have now been recognized as National Parks. The flora, in particular the rare and endangered plant species, is being studied in these National Parks. A doctoral thesis has just been completed on the Kaziranga National Park and Manas Sanctuary. So far there have been no deliberate introduction or re-introduction of rare species in any of the parks or sanctuaries.

India is a signatory to the Convention on International Trade in Endangered Species of Wild Fauna and Flora (CITES). Its suggestions are included in the Appendices of the Convention, and the list is under revision for the next meeting of the Convention in March 1979 in Costa Rica.

Botanic gardens can also play an important rôle in preventing the smuggling of rare species and in discouraging requests for 'prohibited' species. Small populations can be grown to meet the requirements of education and research. The experimental garden of the Botanical Survey of India at Shillong has about a dozen populations of the pitcher plant *Nepenthes khasiana* and pitchers are supplied free to educational institutions.

Apart from the various international or governmental activities on conservation, very strong public opinion and awareness has been created in most parts of the country by private organizations. There are many societies like the Friends of the Trees and the Indian Environmental Society in the country to resist even governmental actions on deforestation. A few years ago, in the Himalayan region, a movement was started by some self-appointed conservationists for the protection of trees and they called this the 'Chipko' movement (in local language 'chipko' means 'to cling to'). The volunteers of this movement embraced trees when any squad of forest contractors wanted to cut down the trees. The volunteers offered to sacrifice their lives along with the trees. This had a great impact and many deforestation programmes had to be stalled. There are reports also of resistance to deforestation in other ways from other parts of the country and in this respect the message of conservationists in the country seems to have reached the masses.

The Botanical Survey of India is trying to contribute its little mite in the mighty task of conserving the rare flora and habitats of the country. This contribution is done through (1) the actual survey, study and evaluation of the status of some individual threatened species and threatened habitats, (2) through supplying data on the flora and its ecology to national and regional committees, societies and research teams, (3) through the constant monitoring of the programmes of MAB, CITES, NCEPC, and IBWL in India, so far as they relate to flora, and (4) finally through advice and education to numerous governmental and private bodies, associations and societies on the need and methods of conservation.

The flora of this earth belongs to the whole world, this whole generation and all future generations. 'We', 'you' and 'they' are all custodians of the limited flora that grows on that piece of land or water which by history and circumstances is now under our administration and care. It is, therefore, my hope and appeal that when we go back home from this conference, we will return more inspired and committed to contribute our rôle in conservation of the flora of this *one earth, the only earth.*

Mexico: Practical Conservation Problems of a New Botanic Garden

A. P. Vovides

Instituto de Investigaciones Sobre Recursos Bióticos, Xalapa, Veracruz, Mexico

INTRODUCTION

Botanic gardens have existed in Mexico since pre-Hispanic times and they have reflected the great love and respect which the ancient Mexicans had for nature. These gardens were probably amongst the first botanic gardens of the world. Mexican emperors created gardens to which they introduced plants and animals from other regions of the country, especially plants of medicinal value; the gardens of Montezuma in Oaxtepec were enriched with tropical medicinal plants that could not be grown on the Mexican 'altiplano'.

With the coming of the Spanish conquest these gardens were destroyed, with the exception of the famous pre-Hispanic garden of Chapultepec which existed as a botanic garden until recent times, and although now of a different character, it can still be seen today as the major park of Mexico City.

Modern botanic gardens in Mexico are few, however. In Tuxtla Gutierrez, Chiapas, a botanic garden was founded some thirty years ago by the renowned Mexican botanist, Dr Faustino Miranda, who also founded the botanic garden at the National University of Mexico in Mexico City, where there is a large open air cactus collection and a display greenhouse of tropical plants landscaped in a unique way. This garden is used mainly for research and student education.

The most recent botanic garden of Mexico was founded on 16 February 1977 at Xalapa, the state capital of Veracruz. The garden is 1,300 metres above sea-level; it lies 122 kilometres from the port of Veracruz and 302 kilometres from Mexico City. The garden is situated on the windward slopes of the Sierra Madre Oriental in a cloud forest environment with high rainfall and humidity accompanied by frequent mists. Annual rainfall is 1,454 mm and the mean annual temperature is 18°C. with a mean annual relative humidity of 78 per cent. The Garden is named after the Jesuit monk Francisco Javier Clavijero, a Mexican historian of the 18th Century who possessed a great fondness for nature and for the Mexican countryside.

THE BEGINNINGS

It could be difficult to justify the need for a botanic garden in newly developing countries like Mexico because of socio-economic and political reasons, particularly at a time when the country is pressed by urgent problems such as food production and the need to provide hospitals, housing and schools. The biological sciences are often poorly developed in such countries, but Mexico is undergoing a scientific and cultural revolution, and more people especially at governmental level are becoming aware of the problems affecting the environment, and are now recognizing that there is a need for botanic gardens to support botanical studies in relation to the whole environment. The application for funds to start a large garden for botany alone would not be approved, but the funding of a multi-disciplinary institute whose aims are to help solve the country's immediate problems is a different matter.

The founding of the Instituto de Investigaciones Sobre Recursos Bióticos (Institute for Research into Biotic Resources) in 1975 was the result of the concern of the Mexican Government to prevent the deterioration of the country's environmental resources. The Institute's main objectives are to deal with problems such as food production and rational land use, which it does by research into new and at times unconventional ideas.

The State Government of Veracruz acquired approximately 60 hectares of land for a recreational park in Xalapa at about the same time that the Institute came into existence, also in Xalapa. Dr Arturo Gómez-Pompa, Director of the Institute, convinced the Governor of Veracruz of the benefit of a botanic garden for Xalapa, where it would be an additional cultural centre as well as a focal point for putting the Institute's ideas across to the general public. The garden would be a public 'window' into the Institute.

The Governor, already convinced of the need for re-afforestation and plant conservation in the State of Veracruz, provided the Institute with seven hectares of the park along with three State-paid labourers and the co-operation of other departments, who also provided some resources and casual labour. The garden had to work to a low budget, make use of what already existed on the site, make the labels and paths as cheaply as possible and justify its existence as soon as possible. Thus the labels are made out of 'Dymo' plastic tape fixed to painted wooden boards; the paths were laid manually and surfaced with volcanic ash to give a non-slip surface; steps were made out of locally available wood and of old reject wooden railway sleepers donated to the garden by the Mexican National Railway Company.

Skilled garden staff are not available in Mexico since colleges do not offer horticultural training; horticulture does not exist as a formal trade or profession. We rely on unskilled 'campesino' labour and at best self-taught semi-skilled gardeners. All work is carried out manually because machinery is very expensive; thus a large labour force is required for major projects. We hope to partially resolve this problem in the near future by the purchase of essential equipment including a tractor.

We were fortunate that the botanic garden is situated on a wooded hillside, rich in *Quercus* spp., *Liquidambar* and *Carpinus*, the remnants of a former forest. These trees have been identified and labelled together with the shrubs and herbaceous plants. We are taking advantage of this natural wooded hillside and are constantly enriching it with native forest tree species and other regional plants, especially epiphytes such as bromeliads and orchids. The hillside has a series of ash paths, cut through it at low cost, which disturbs the habitat as little as possible and enables the public to see the plants in the way they would in the wild. We add the local common names and uses to the normal botanic garden system of including the family, genus, species and geographic distribution on the plant labels. This wooded hillside was open to the public within six months of work beginning on the garden and the great age of some of the trees helps give this part of the garden an appearance of long establishment.

Response from the general public was slow at first, and the type of people who first came to the garden already knew something about botanic gardens or were keen naturalists, so that it was generally a case of 'preaching to the converted'. Schools were reluctant to visit us in spite of the publicity directed at them. We put this down to the teachers not appreciating the usefulness of a botanic garden as a teaching aid. School visits increased after audio-visual talks had been given to teachers in the schools followed by special guided tours to back this up; subsequently many schools visited us and in turn teachers recommended the garden to their colleagues so that now we have a steady flow of visiting schoolchildren. Our aim must be to educate the next generation, since the older generation will not change its attitude towards conservation very easily, and is unlikely to respond directly to a botanic garden and its objectives.

The main aims of the garden are to display the regional flora; to introduce useful wild plants and ornamentals from similar regions of Mexico; to be an introduction centre for tropical montane plant species; to initiate public interest in the care of and respect for wildlife generally; to maintain small groups of rare, threatened or endangered regional plant species for protection and propagation purposes; to further the propagation of native forest trees for use in local re-afforestation programmes; to demonstrate good land use through vegetable gardening, terracing and composting; to be an educational centre for all levels, from kindergarten to post-graduate; to display the art of good gardening through a small formal area consisting of native and exotic ornamentals landscaped for aesthetic reasons and finally to have live plant material on hand for scientific research.

Traditionally botanic gardens are attached to universities or are institutions whose primary objectives are scientific research and student education. In Veracruz we have to justify our existence in the eyes of the government and the public. Our activities are therefore tailored to tackle the current problems facing the country. Thus we have inverted the traditional priorities of first scientific research, then education, then conservation into education as our main priority, conservation through education as an immediate second priority, then scientific research. In this way we are making our presence felt and the public is becoming more involved and interested in conservation matters through actual contact with the living plants. Individuals and local authorities are now putting to us the problems they face concerning plant conservation. We have already donated thousands of native tree seedlings to municipal and other authorities for re-afforestation projects. These trees are produced in our small nursery which is maintained at low cost by using naturally occurring resources such as bracken leaves for shading.

A good example of rational land use is demonstrated by a vegetable garden stocked with a variety of locally well known vegetables as well as unconventional ones for the region; a modified version of the floating garden ('chinampa') is employed. This technique was used by the Aztecs during pre-Hispanic times for producing maize and other vegetables for household and market use. The method still survives today, although on a much smaller scale, in the Xochimilco district, mainly for the production of flowers and ornamental plants. The mud, rich in organic matter from the bottom of swamps, is used to germinate the seeds in special beds ('almácigos'). The seedlings are then transplanted from these beds in a small block of mud ('chapín'), similar to a peat pot, to their final positions. Though this system is labour intensive, it is highly productive and no expensive technology or equipment is required, thus making it readily adaptable to the 'campesino' way of life. This point was proved beyond doubt by the Institute during 1976–77 when this method of cultivation was introduced into the lowland tropical swamps of Tabasco, where the 'campesino' population has virtually no employment during the wet season. The system enabled them to take advantage of the organic rich swampy environment to produce vegetables. Owing to its success elsewhere the Garden feels that a well publicized demonstration of this method is needed to enable the 'campesinos' of Veracruz to take advantage of it.

THE COLLECTIONS AND ENDANGERED SPECIES

The lack of an inventory of the flora of Mexico coupled with the daily destruction of whole habitats such as lowland tropical rain forest, montane rain forest and cloud forest makes the assembly of an endangered species list exceedingly difficult (Vovides & Gómez-Pompa, 1977). Although we are not obsessively interested in individual endangered species, we are very concerned with the conservation of whole ecosystems which contain such plants. Although botanic gardens can in no way substitute living plant collections for conservation in the wild (Budowski, 1976), botanic gardens can play an important rôle in public education as well as being 'lifeboats' in the maintenance and propagation of endangered species and could sometimes become the only source for re-introduction of these species into selected reserve areas. This has been done with certain animal species, so why not with plants? I feel that the collaboration of conscientious commercial nurseries could also be of great value in this respect.

We try to collect from the least disturbed areas in order to obtain a good representation of the native flora, although this is not always possible as these last remaining localities are often both inaccessible and dangerous, for example the sides of ravines. We also collect orchids and other epiphytes from areas that are being clear-felled, and we earmark areas of outstanding interest as possible future reserves. These are reported to the Dirección de Asuntos Ecológicos, a new State Government office concerned with ecological matters, where the areas are investigated with the possible intervention of the State for their purchase and conservation. Areas already listed for conservation are: Cerro Nanchital, lowland tropical rain forest in the Rio Uxpanapa region; Sierra de Santa Marta, tropical rain forest in the Los Tuxtla area; Sierra de Chiconquiaco, cloud forest near Xalapa; and the Cofre de Perote, oak and pine woodland above Xalapa.

We have some well known Mexican orchids in our collection such as *Laelia anceps*, *Brassia verrucosa*, *Odontoglossum rossii* and *Lycaste deppei*, and we are beginning a programme of propagation of endangered orchids by means of seed and meristem culture. An example of this is the propagation of *Lycaste skinneri* var. *alba*, a plant worthy of special mention. We obtained seed from a private collector who acquired this plant from Chiapas some years ago, and although according to orchidologists this species is now probably extinct in Chiapas, it still occurs in Guatemala where over-collecting and habitat destruction are daily reducing its populations.

Tree-ferns were once a common feature of the Xalapa countryside but now they are becoming less common owing to forest clearance and the over-exploitation of the 'trunks' to make fibre pots and holders for orchid growing and for local handicrafts. These plants are becoming vulnerable and some species are even endangered. The Institute has published a leaflet on the tree ferns explaining to the public their sad plight in the hope that this will help stop their decline. Local tree ferns that we have on display include *Sphaeropteris horrida*, *Cyathea fulva*, *Nephelea mexicana* and *Trichipteris bicrenata* amongst others.

The cycads are another important group of plants in our region that need protection. We are attempting to build up a collection of all the Mexican cycads for propagation and education purposes. Viable seeds have been collected in the field as well as live plants; at present we have *Dioon edule, Zamia fischeri* and *Ceratozamia mexicana* seedlings and we are also interested in other forms of propagation of this relict but interesting group of plants.

Amongst other local endangered plants we have *Hydrangea nebulicola* from the Sierra de Chiconquiaco, but this species is proving difficult to establish in the garden. Another interesting plant is *Symplocos coccinea*, again somewhat difficult to propagate but we hope to introduce this into the horticultural trade because of its long-lasting perfumed flowers. Many native magnolias are now becoming scarce owing to habitat destruction. Another interesting plant which is threatened is *Talauma mexicana*; its large fleshy flowers are quite distinctive and were known to the Aztecs as 'yoloxóchitl', which means 'heart-flower'. It is used up till this day as a medicinal plant and delicacy, and the dried flowers can be bought in some local markets, especially in Córdoba, Veracruz. Another species now believed to be extinct in the wild is *Zea (Euchlaena) perennis*; cuttings of this were obtained from Dr Iltis of the University of Wisconsin about a year ago, and we are now propagating it to distribute to other gardens and institutes.

PROMOTION OF THE GARDEN

During the first year of the garden's existence we undertook promotional activities to attract the general public to the garden. Since the great majority of the Mexican people are unfamiliar with botanic gardens, we needed to introduce the garden to them through a series of activities, the idea being to attract them to the garden for a local event in the hope that afterwards they would take the trouble to look round the grounds. We offered special guided tours to encourage this. The year 1977 saw a number of these events, such as a flower show in May (the first formal flower show for Xalapa) and a series of musical events during the month of August, which included chamber music, jazz, latin-american folk music and a 'ballet folklórico'. In October of the same year we laid on a puppet show aimed mainly at the local primary schools.

THE FUTURE

Thanks to the success of the garden, the State Government has given us forty hectares of land designated as a recreational park, adjacent to the botanic garden. Our first objective will be a nature trail through unspoilt forest.

We are interested in collaborating with other gardens and this year Mr R. I. Beyer, Deputy Curator of the Royal Botanic Gardens, Kew, helped us by advising on future planning and development. We are looking forward to a further close liaison with Kew in this respect.

CONCLUSIONS

Botanic gardens are urgently needed in the tropics. This was stressed in the past by Professor E. J. H. Corner: "The number of Botanic Gardens in the tropics should be increased and should develop better relations, particularly in the loan or exchange of staff" (Corner, 1946). Recently Dr Melville (1976) mentioned that there was a need for tropical botanic gardens to undertake conservation projects.

It is our hope, through justifying the existence and need for our botanic garden, that it will be the first of a series of gardens of this type to be initiated in other important regions of Mexico.

ACKNOWLEDGEMENTS

I am most grateful to the following for the valuable advice and criticism they have given me in the preparation of this paper: Dr Arturo Gómez-Pompa, Dr Ramón Echenique, the members of the Cambridge Mexican Botanical Expedition 1978, and Dr John D. Rees for the help and advice on the collection and cultivation of the cycads. I am especially grateful to Mr R. I. Beyer of the Royal Botanic Gardens, Kew, for the valuable advice we received during his stay at Xalapa during 1978.

REFERENCES

BUDOWSKI, G. (1976). The Global Problems of Conservation and the Potential Role of Living Collections. *In 'Conservation of Threatened Plants'* (eds J. B. Simmons *et al.*). Plenum Press, New York and London. Pp. 9–13.

CORNER, E. J. H.(1946). Suggestions for botanical progress. *New Phytol.* **45(2)**: 185–192.
MELVILLE, R. (1976). In Discussion, Sect VI: Techniques of Collecting. *In 'Conservation of Threatened Plants'* (eds J. B. Simmons *et al.*). Plenum Press, New York and London. p. 182.
VOVIDES, A. P. & A. GÓMEZ-POMPA (1977). The Problems of Threatened and Endangered Plant Species of Mexico. *In 'Extinction Is Forever'* (eds G. T. Prance & T. S. Elias). New York Botanical Garden. Pp. 77–88.

POSTSCRIPT (JUNE 1979)

A new location for *Zea perennis* (see p. 121) was found in Jalisco in December 1977, by members of the Agricultural School of the University of Guadalajara.

South Africa: The Conservation Policy of the National Botanic Gardens and its Regional Gardens*

A. V. Hall   &   H. B. Rycroft

University of   Kirstenbosch Botanic
Cape Town   Garden, Cape Town

INTRODUCTION

Since their foundation in 1913, the National Botanic Gardens of South Africa have actively promoted public interest in appreciating, cultivating and conserving wild plants. This has been aided by a close association with the Botanical Society of South Africa, a voluntary body with a large membership, with its headquarters at the main Gardens at Kirstenbosch near Cape Town.

This policy of promoting public interest has been followed in a variety of ways. On the premise that the best way to win co-operation is to have participation, the Gardens have strongly encouraged the growing of South African plants in private gardens and public parks. Many thousands of packets of seeds are distributed each year for this purpose. This is backed by the Botanical Society's mailings of a general-interest journal, *Veld & Flora*, together with literature and book-lists on cultivation and conservation. For schoolchildren, the Kirstenbosch Gardens are an ideal area for nature study and a centre opened for this purpose in the 1920s is now a flourishing concern. This has been most valuable in stimulating public concern for wild plants as well as their conservation (Compton, 1965). For the informed layman and the scientist the Gardens provide an herbarium and library, and publish a quarterly scientific periodical, the *Journal of South African Botany*. Close links are kept with the University of Cape Town and full use is made of the Kirstenbosch Gardens by students and staff for study and research. The Director of the Gardens holds a professorial post in the University.

The Kirstenbosch Gardens lie at the foot of the steep, eastern slopes and gorges of Table Mountain. These areas are covered with natural forest, silver-tree woodland, and heath and protea veld. The Gardens merge gradually into this wild vegetation. This has made possible the unusual but highly successful policy of keeping entirely to South African and chiefly Cape plants in the Gardens. Numbers of rare plants, including the Silver Tree, *Leucadendron argenteum* (Proteaceae), grow naturally under these conditions. The mountain backdrop gives a strong reminder of the importance of conserving the wild habitats of plants in the rich and beautiful scenery of the south and southwestern Cape Province. Regional Gardens are being established in other centres across South Africa, wherever possible similarly linked to attractive wild scenery.

* Read by H. B. Rycroft

125

Several of these Regional Gardens are already flourishing concerns, encouraging pride of place and interest in the local flora. Each of the nine gardens devotes its entire attention to the native flora of its own local area.

This policy of the encouragement of public interest in the South African flora is of the utmost importance for conservation. Over the vast territories of Southern Africa the conservation of rare plants in the wild very often rests solely on private initiative. Agriculture is today one of the major threats to rare plants in the region. Farmers have to provide food and other products for a population that is increasing by 60,000 persons a month. It is generally the farmer's decision whether to plough up an area of wild vegetation, to introduce grazing, or to spend hard-earned funds on suppressing invasions by alien plants that smother the indigenous flora. The reservoir of goodwill shared by land-owners and farmers is a first line of defence in conserving many of South Africa's rare plants. It is clear that the Gardens' policy of interesting the public in wild plants must continue with all possible strength. However, the Gardens also play a valuable rôle in directly conserving threatened and rare plants.

THE NEED FOR A POLICY ON THREATENED AND RARE PLANTS

A definite policy towards threatened and rare plants has not yet been defined by the Gardens, chiefly because of a lack of a survey of the scale and importance of the problem in Southern Africa. This lack of evidence is being remedied. A list of threatened and critically rare plants in Southern Africa has been in preparation since 1974 and will shortly be published (Hall *et al.*, in press). Detailed field studies of threatened populations are in hand. An intensive survey has been mounted in the greater part of the Cape Province. Here, in an area about the size of France, there are 38 species presumed recently extinct, 68 Endangered, 84 Vulnerable, 278 Rare, 218 Indeterminate, and 773 uncertain but suspected to be in hazard (Hall, 1978a). Figure 1 shows how many of the threatened and rare plants are concentrated in the coastal mountains and flats of the south and southwest. The flora here is extraordinarily rich in distinctive taxa and, because of this, forms on its own one of the world's six Floristic Kingdoms (Good, 1964). About 60 per cent of the area for the Cape Kingdom has been taken over by agriculture, forestry and urban development. This has reduced it to isolated patches and narrow mountain corridors that together are about the size of the Kruger National Park (Hall, 1978a). The remaining wild areas have long boundaries of contact with human pressures, such as invasions by thicket-forming introduced alien weeds, increased frequencies of fires and destruction of pollinators by insecticides.

The reduction of wild areas may have had an additional effect besides making species threatened or extinct. By destroying local populations of more widespread species, significant segments of species-variation may have been lost. The base for future evolution in Southern Africa may have been as seriously impoverished by this as through the loss of variation recorded in the lists of extinct and threatened species. The biological and the human-based reasons for halting this

FIGURE 1. Map showing the numbers of threatened and rare species in 1×1 degree latitude and longitude areas in the Cape Province, South Africa.

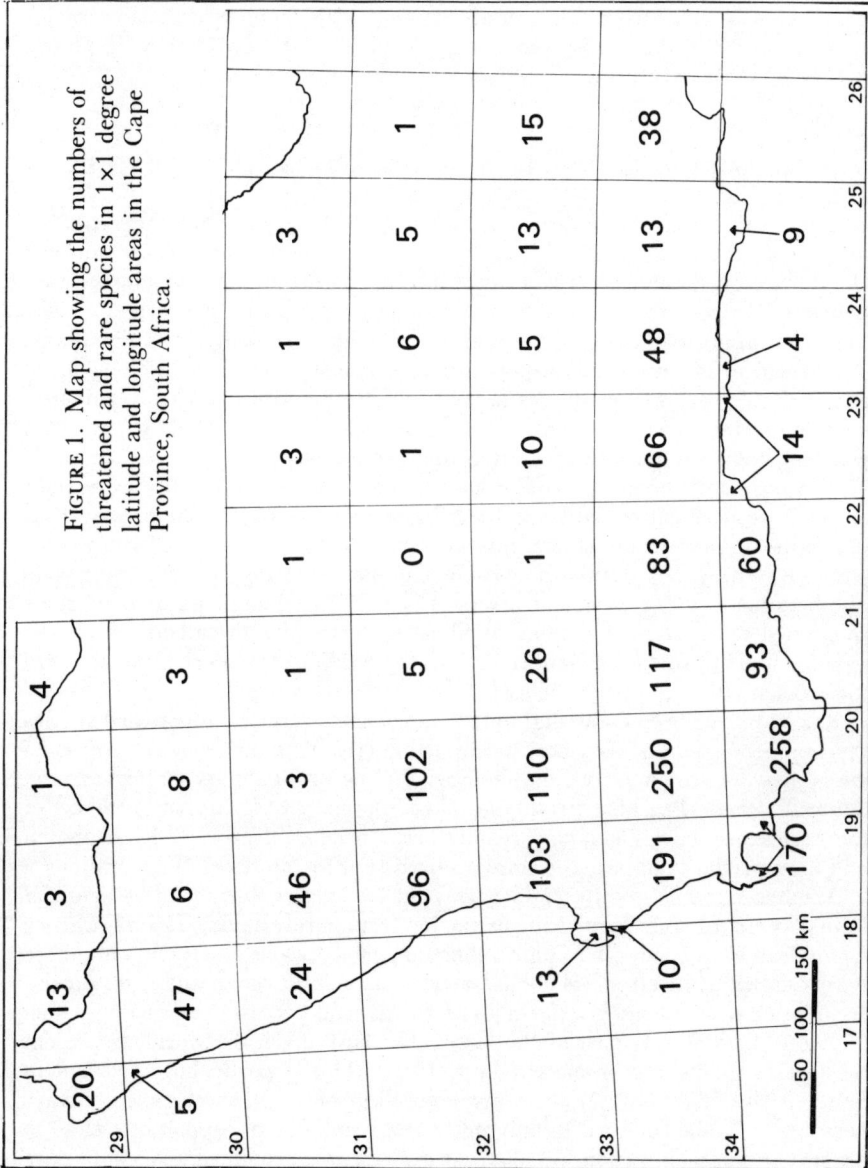

tide of destruction are given in detail with the threatened and rare plant list for Southern Africa (Hall *et al.*, in press).

The problem is clearly of a magnitude and importance that demands urgent action. From what is described below it is evident that the National Botanic Gardens of South Africa can usefully develop both their existing concern for conservation, and their long-standing horticultural expertise with wild plants, to help meet this problem.

ESSENTIAL THEORETICAL ASPECTS STILL NEEDING FURTHER STUDY

Several basic aspects of the threatened plant problem still remain under-studied in spite of the advances recorded at the previous conference in this series. Lack of evidence on these aspects is a hindrance to the formulation of conservation policies.

Of critical importance to botanic gardens is to know broadly how many plants are needed to make a survival-population genetically viable in the long term, and how this varies from species to species. There is still barely any guidance on this crucial question (Mayr, 1963; Terborgh, 1974; Frankel, 1976). This problem stems from a lack of study of the process of extinction from the point of view of population genetics. This gap in knowledge is remarkable in the light of human-caused extinction being the 'ultimate sin' for conservation, as well as the serious consequences of the present increased rate of loss of plants world-wide (Raven, 1976). Until the problem of not knowing the minimum safe numbers for carrying sufficient diversity is resolved, botanic gardens must be committed to a policy of holding as much material of a threatened species as possible without further endangering it in the wild, if indeed practical steps are to be taken with the plants themselves.

It is encouraging to know that in the case of some rare and threatened species the number of plants does not appear to be critical in the shorter term even when they are grown away from their original natural habitats. *Protea aristata*, *Serruria florida* (Blushing Bride) and *Leucospermum reflexum*, all members of the Proteaceae, have taken very readily to artificial cultivation at Kirstenbosch and have produced apparently unhybridized seed for numbers of generations.

Another aspect is to differentiate between the species threatened by artificial human pressures and those facing extinction from purely natural factors. Theory would lead us to believe that, unless there are good reasons otherwise, only those plants under artificial stress should receive deliberate conservation treatment. By this one would avoid interfering with the natural process of extinction which has been a prominent part of the forces that have given the world its present heritage of plants and animals (Mayr, 1963). This is particularly relevant in South Africa, especially at the Cape. Numbers of Cape species have always been known to be confined to only one or two small areas, apparently close to a state of natural extinction. An example is *Erica fairii*, confined to a one-hectare patch on a plateau near Cape Town. This localization in the Cape flora may be due partly to post-Pleistocene arid conditions shrinking the area for mesic plants into the moist coastal belt, with a consequent intensification of competitive pressures (Hall & Boucher, 1977).

It is already often difficult to differentiate between natural and human-based extinction. Large and fully pristine plant and animal communities are becoming increasingly rare in Southern Africa. Many appear to have been affected by little-studied, cryptic, but nevertheless serious changes such as the destruction of pollinators or seed dispersers, new grazing patterns, or altered burning regimes. With this blurring of the picture of natural extinction, it may be best to rescue as many species as possible. Certainly with the loss of diversity that has already occurred it is tempting to try to conserve all populations that may seem to be threatened.

PRACTICAL ASPECTS

With this background, the development of a conservation policy for threatened and rare plants by the National Botanic Gardens of South Africa presents both challenges and ready opportunities.

One may assume from theory that the populations for long-term survival should ideally be located in the original natural ecosystem. It is here that they can continue to experience the ancient amenities and pressures that have caused them to survive over many past generations, and which, by inference, would allow them to last well into the distant future.

Where conditions are much less than ideal it may be impossible to find even a quasi-original habitat that could be restored and given some degree of safety. A habitat consisting of lowland relicts in the wheat-growing country north of Cape Town was recently found to have previously shrunk from about 600,000 hectares to 4,500. These relicts are in broken patches severely under pressure from ploughing and from invading thickets of Australian *Acacia saligna* and other alien plants. The largest patch was 750 hectares in area, and contained 28 threatened and rare plants, some only known from this site (Hall, 1978b). Even with intensive conservation, the long-term future of these relicts may be threatened by cryptic pressures. Essential populations of pollinators may be damaged by the drift of insecticides from aerial spraying of adjoining farm-land. Salinities may rise on the flats due to the leaching of fertilizer salts from surrounding soils. Field studies are showing numerous cases where the habitat and the plant it contains are both almost or fully destroyed. *Gladiolus aureus*, now down to 35 plants in the wild, exists at a site under pressures from flower-pickers, trampling by playing children, picnickers, drainage alteration and invasive alien weeds. *Moraea loubseri*, discovered as a single population in 1973 on a granite hill-top, has had its habitat destroyed by quarrying and is now probably extinct in the wild. Both species are of horticultural interest and small populations have been rescued and are in cultivation at the Kirstenbosch Gardens and elsewhere.

The horticultural aid that the National Botanic Gardens are giving varies from restoring depleted wild populations, to giving a temporary home for evacuees before they are taken back to new and safer natural areas.

Horticultural restoration in the wild would involve assisting the threatened plant in producing more individuals by pollinating all flowers, protecting seeds and seedlings from predators and perhaps also raising young plants in nurseries. This needs to be done in close co-operation with the bodies that are charged with proclaiming new nature reserves. These bodies, chiefly the Provincial nature conservation departments and the Department of Forestry, would also be responsible for vegetation-management. This would be primarily concerned with bringing artificial pressures under control in the threatened plant's habitat. Former natural factors would need to be restored. An interesting example is that fire is known to be essential for the completion of the life cycle of the Marsh Rose (*Orothamnus zeyheri* of the Proteaceae), which as a result is now well on the way to recovery from a severely endangered status as controlled burning is included in the management regime. Horticultural restoration work would require an increase in the Gardens' staff and facilities.

Where the habitat is beyond restoration, some other safe wild place must be found and the species transferred there. If the former habitat is facing imminent destruction, or where the plant will take some time to become convincingly established at a new site, a temporary survival-population may have to be kept in a nursery in the botanic garden. As noted above, the minimum safe size of the survival-population is not known. Another problem is that to avoid effects of unknown selective pressures or hybridization in horticulture, the plant must be allowed to pass through as few generations as possible in the gardens. This difficulty may be especially acute for annuals. Where two or more populations are sampled to provide the survival-set there may be genetic advantages in retaining distinctness of each in cultivation by keeping them apart so as to control gene exchange. This could present difficulties in the confines of a botanic garden.

Besides the plants that are kept for providing propagules for testing possible new habitats, a large segment of the survival-population could be held in the form of seed in a refrigerated seed bank. For many species, this may offer great security. Cold storage of the seeds of critically endangered species has been started at the University of Cape Town. The seeds have been collected in the wild, cleaned, counted and stored over anhydrous silica gel at 3°C. for four weeks prior to being transferred to storage at −22°C. Moves are being made by the National Programme for Environmental Sciences towards the establishment of a seed-testing laboratory. This will allow the essential periodic testing of the viability of the stored seed.

Assessing where the new wild sites for evacuees should be located will be a matter of great skill. There will need to be close co-operation between the horticulturist and the ecologist. Any translocations, whether immediately successful or not, should be recorded in the scientific literature. The new sites should be as close as possible to the old, without prejudicing either the safety of the species or the standards of the ecological amenities and natural pressures. *Moraea loubseri* may be established on a chemically and physically similar granitic soil to that at its former site, in a nearby nature reserve. *Gladiolus aureus* will have to be re-established some distance from its former site, at the nearest safe place where similar soils and microclimates appear. Having the new and old sites relatively close may be important where the plant must rely on a specific local pollinator.

Other obscure factors may obstruct the integration of a rare plant into a new community. Care must be taken to avoid places where related species might prejudice the future of the evacuee by excessive competition or by hybridization. In all this, it must be remembered that the need to establish even a temporarily safe wild population may be too urgent to await detailed research. In some cases, the policy for action will have to be largely based on a mixture of shrewd observation and inspired guesswork, based on the traditions established with related plants.

Some plants may prove difficult or even impossible to re-establish in the wild, although they may survive quite well in the protected, special conditions of the botanic garden. It may be reasonable to depend upon their survival in horticulture if they are attractive or interesting. For others, it would be too much to expect of a botanic garden to ensure that they would continue for long periods in cultivation, simply because they are threatened in the wild. An alternative to extinction may be the storage of their seed in a seed bank. At least this would hold open the options of growing the plant for future human or biological use, or re-establishing it in the wild when ecosystem dynamics are better understood.

The number of species for which horticultural aid would be useful in rescuing them from endangerment may be estimated at two or three hundred in Southern Africa. A more precise estimate will be needed before proper planning can take place. It is still too early to provide this. Surveys of the actual status of the populations of threatened species are proceeding slowly and completed field studies are available for only about 5 per cent of the nearly 2,000 species listed in the report on the threatened and critically rare plants of Southern Africa (Hall *et al.*, in press). Horticultural aids to rescuing threatened plants have been run on a small scale for many years and it would seem reasonable to expand this activity gradually as new data become available on likely candidates for assistance. Gardens staff are already being aided in exploratory work by listings from the computer-indexed data bank on threatened plants at the University of Cape Town (Hall, in press).

A programme of this size and difficulty will require an abundance of professional and technical skill. By the very nature of their small numbers in the wild, rare plants may often prove especially intractable to horticultural methods. The programme could not be adequately run by the present staff of the National Botanic Gardens alone, who are already actively overloaded. New professional and technical posts will be needed, with good salaries to retain staff of a high calibre. These persons would work under challenging conditions, especially where their skills would be applied away from the controlled habitats of the gardens, to restore, strengthen and even establish populations of plants in the wild. The posts would need to be distributed in a way that would make maximum use of the special climatic opportunities offered by the Regional Gardens.

Creating this practical programme will require both inspiration and firm leadership. The inspiration can come from an understanding that humanity now has a large share in upsetting the balance of nature. As part of the system, humanity has an implied responsibility to help create a corrective response, in this case the restoration of an outstanding heritage of plant life.

SUMMARY AND CONCLUSIONS

It is concluded that besides its well established programme of stimulating an interest in wild plants, the National Botanic Gardens of South Africa could play an important rôle in co-operative programmes to help conserve threatened species in nature. Horticultural aid may be required for an estimated two or three hundred of the most threatened plants in the South African flora. Both restoration and evacuation away from threatened habitats may be needed for wild populations of plants. Difficulties in this work are outlined. The most notable is a lack of guidance on the size and variability of the minimum survival-population of different species.

POSTSCRIPT

Although this paper concerns conservation in South Africa, Professor Rycroft in his personal capacity is also concerned with species conservation in a global sense. With his wife he has a farm near Stellenbosch 65 km from Cape Town, where they have set aside 200 hectares of native flora as a private nature reserve, proclaimed by the Provincial Government as the 'Brian Rycroft Nature Reserve', and there they have commenced the establishment of an arboretum and garden for plants from all parts of the world.

The area has a warm-temperate. winter-rainfall climate but water, which is readily available, can be applied at any time of the year. They invite anyone interested in conserving any species by cultivation to send propagating material in the form of seeds or otherwise and they will do their best to co-operate.

Furthermore arrangements can also be made in Kirstenbosch, or in fact in any of the Regional Botanic Gardens spread throughout South Africa, to accept and attempt to propagate and hold any species that readers may care to send for trial.

DISCUSSION

C. D. ADAMS (Trinidad), noting that a large part of the Kirstenbosch Botanic Garden was apparently natural vegetation, asked if a policy of burning these areas was being implemented, in order to protect the rare species that required periodic firing.

H. B. RYCROFT replied that firing had taken place accidentally about every 10 to 12 years, and so a firing policy as such was not required. Furthermore a gap of 8–15 years between firing had not been sufficient, but if the vegetation was left unfired for up to 20 years some species of *Protea* tended to die out.

C. D. ADAMS added that in other parts of the world where areas required periodic burning, a long spell without a fire was detrimental, since when the fire eventually occurred the intense heat tended to destroy much of the woody vegetation.

H. B. Rycroft agreed and commented that in some areas of 'tough' grassland there were fires every year. Furthermore, certain native species would only flower for two or three years following a fire. When developing a firing policy many variables had to be considered, such as the type of vegetation, the climate, the current weather, the time of the last fire, etc.

J. P. M. Brenan (Kew), having complimented the speaker on his approach to public relations at Kirstenbosch, enquired how the programme of removal of *Eucalyptus* had proceeded.

H. B. Rycroft explained that until recently there had been the problem of alien species, particularly *Eucalyptus*, invading some 80 per cent of the Kirstenbosch garden. With repeated removal of these aliens this had been reduced to 2 per cent and native species had replaced the alien plants. However, much work was still required to maintain this improved situation.

A. P. Vovides (Mexico) requested further details on the programme of public education at Kirstenbosch.

H. B. Rycroft said two full-time school-teachers were employed; each took a class for a whole day, instruction mostly being given outdoors in the gardens. Classes visited the garden about four times a year; in some cases attendance was for one hour a week for a ten week course. Another means of stimulating children's enthusiasm was through essays and pictorial competitions, which attracted large numbers of applicants. There was also a system of trained honorary guides who manned the garden offices, etc., when permanent staff were away at week-ends, took parties around the garden and so helped to stimulate interest amongst the public.

W. Richter (FRG) asked whether everyone could receive instruction when visiting the botanic gardens of South Africa.

H. B. Rycroft stated that all races were welcome.

REFERENCES

Compton, R. H. (1965). *Kirstenbosch, Garden for a nation*. Tafelberg-Uitgewers, Cape Town. 168 pp.

Frankel, O. (1976). The Time Scale of Concern. In *'Conservation of Threatened Plants'* (eds J. B. Simmons *et al.*). Plenum Press, New York and London. Pp. 245–248.

Good, R. (1964). *The geography of the Flowering Plants*, 3rd Ed. Longmans, London. 518 pp.

Hall, A. V. (1978a). Endangered species in a rising tide of human population growth. *Trans. Roy. Soc. S. Afr.* **43:** 37–49.

—— (1978b). *Riverlands Fynbos habitat*. Threatened habitat report. University of Cape Town, unpublished typescript. 8 pp., 7 appendices.

—— (in press). Information handling for South Africa's rare and endangered species survey. In *'Proceedings of a Conference on Geographical Data Handling for Rare Plant Conservation'* (ed. L. E. Morse). New York Botanical Garden.

HALL, A. V. & C. H. BOUCHER (1977). The threat posed by alien weeds to the Cape Flora. *In 'Proceedings of the Second National Weeds Conference of South Africa'*. Balkema, Cape Town. Pp. 35–45.

HALL, A. V., M. de WINTER, B. de WINTER & S. A. M. van OOSTERHOUT (in press). *Threatened and critically rare plants of Southern Africa*. South African National Programmes Report. Council for Scientific and Industrial Research, Pretoria.

MAYR, E. (1963). *Animal species and evolution*. Harvard University Press, Cambridge, Mass. 797 pp.

RAVEN, P. H. (1976). Ethics and Attitudes. *In 'Conservation of Threatened Plants'* (eds J. B. Simmons *et al.*). Plenum Press, New York and London. Pp. 155–179.

SIMMONS, J. B., R. I. BEYER, P. E. BRANDHAM, G. LL. LUCAS & V. T. H. PARRY (eds) (1976). *Conservation of Threatened Plants*. Plenum Press, New York and London. xvi+336 pp.

TERBORGH, J. (1974). Preservation of natural diversity: the problem of extinction prone species. *Bioscience* **24**: 715–722.

United Kingdom: A Phytosociological Layout for Locally Endangered Species

E. E. KEMP

University Botanic Garden, Dundee, Scotland

In the autumn of 1971, the author was given the opportunity of creating a Botanic Garden for the University of Dundee. The main function of the Garden is to supply plant material to the Department of Biological Sciences for teaching and also to provide accommodation for populations of plants grown for research.

The advantage of the site is its proximity to the Department, approximately two kilometres distant, and it was the last vacant piece of land available in the locality. The area of the Garden is approximately 10 hectares and was mainly neglected arable land which had borne heavy crops of weeds, the seeds of which were unfortunately ploughed into the soil. As a result, whenever there is disturbance during planting operations, a dense sward of weed seedlings appears.

In planning the Garden, it was considered that a plant sociological unit would enhance its educational value at all visitor levels—academic, school and lay public. Although originally conceived only as an educational feature, two additional functions soon became apparent. The huge volume of teaching material required from the countryside suggested, in the interests of nature conservation, that collecting of most of the specimens be from within the plant sociological unit, a means of production less labour dependent than the usual horticultural nursery methods. The other function, the need to provide ecologically appropriate accommodation for locally endangered species, was suggested by the recent publicity from the previous Kew Conservation Conference (Simmons et al., 1976) regarding the rôle which botanic gardens could fulfil in preserving stocks of these plants; this function was further emphasized by the numbers of rarities in the region which were available for rescue from civil engineering and forestry operations.

After a lapse of seven years, the following objectives are being achieved by the unit:

1. The native plant material required for teaching is being produced more efficiently than by the usual horticultural methods, especially now that the component plant associations are beginning to become durably established.

2. Arranging the plants in their natural associations and placing these in a naturally occurring sequence, in the present instance from sea-level to mountain top, is serving an ecologically educational purpose. Such a demonstration is not available in nature without travelling considerable distances. It is not suggested that this layout is a substitute for field study, but that it

135

is a time-saving introduction to local ecology and a more effective one than viewing museum dioramas or projected slides.

3. It is providing accommodation for rarities in their appropriate plant associations, which is essential to those species dependent upon others for survival. The inclusion of the rarities is creating an interest in conservation and a teaching aid for this subject. It is not suggested that the preservation of plants in the Garden is a substitute for conservation in the field, but that preservation is preferable to extinction. Furthermore, the availability of cultivated stocks of these rarities relieves pressure upon them in the field by compulsive plant collectors. In Britain, for example, commercially available material has undoubtedly saved the remaining naturally occurring plants of *Salix lanata* and distracted collectors from the Arran *Sorbus* spp. (*S. arranensis* and *S. pseudofennica*) and possibly also from *S. lancastriensis* and *S. bristoliensis*. Scions of these *Sorbus* taxa are available from the Dundee Garden and seeds will be collected on request.

4. Information on the time required to establish *de novo* the various plant associations and experience in their management may have an application in countryside rehabilitation work.

5. The synecological study of the developing associations also seems to offer the possibility of fruitful research.

6. The unit is attracting widespread public attention, promoting an interest in native plants, and proving a refreshing change from a taxonomically arranged layout, common in many botanic gardens, which inevitably results in some plants being environmentally misplaced.

The phytosociological unit is sited in the centre of the Garden where the natural slope of the ground is from three to five degrees, with a southern aspect. The soil in this area is derived from terrace deposits of sand and gravels and is very freely drained. The lack of water for the streams and pools proved a serious disadvantage, and in the first few months of construction it was even accepted that it would be necessary to re-circulate (by pumping) water from the city supply, despite the high concentration of salts which the continuously re-circulated water would contain. However, a spring was found on the site and later two others were discovered.

The layout begins with a pool (eutrophic) at the bottom of the slope at the south boundary of the site. Nearby, there are the beginnings of sand dune plant communities containing *Elymus*, *Ammophila*, *Salix repens* subsp. *argentea*, *Hippophaë* and *Rosa rubiginosa*. From the pool, a path leads northward up the slope, through Scottish woodland types, Oakwood on the right and Ashwood on the left, in which some of the shrub and field layer species are already established. As we ascend, these woods give way to Scots Pine and Birch and beyond these there is a mound symbolic of a Scottish mountain. At the summit, below a semi-circular ridge, a lochan (tarn) has been constructed and from it the stream begins, tumbling down the hillside in a series of waterfalls from small pools (oligotrophic). In nature, the colonization of bare ground usually begins with the lower plants, the algae, lichens and mosses, followed by the herbs, then shrubs, and finally the trees. In order to accelerate the establishment of woodland in the unit, the trees were introduced first, and since large trees are necessary for the creation of suitable environments for the woodland associations, 5 to 7 metre tall specimens of Scots Pine and Oak were planted. In the case of the quicker growing Birch, the specimens were about 3 to 4 metres tall. The trees are all of local provenance with the exception of *Betula pubescens* subsp. *pubescens*. Pure stands of this subspecies which would provide suitable specimens for transplanting were not available locally, but trees were obtained from the Isle of Bute.

The first group of Scots Pine, 6 to 7 metre tall trees from one of the remaining fragments of native Scots Pine forest, the Black Wood of Rannoch, was planted in March 1972. In the spring of 1973, seventy-five 5 to 7 metre tall Scots Pines of north-eastern Scottish provenance were planted, with the result that the Scottish Pinewood is the most established of the woodland types, with well stocked shrub and field layers. Forest floor litter for the species dependent upon it, such as *Goodyera repens*, was introduced. Seedlings of *Juniperus communis, Vaccinium vitis-idaea* and *Empetrum* have sprung up; *Vaccinium myrtillus* and *Calluna* are well established and also *Goodyera, Trientalis* and *Pyrola media*. Many fungi have appeared from the pinewood floor litter and among the higher plants inadvertently introduced with the balls of soil attached to the trees are *Genista anglica, Galium saxatile, Viola riviniana* and *Oxalis acetosella*, and the pines themselves have begun to regenerate. Other units becoming established are a fern-rich juniper scrub association and a dwarf shrub heath which partly clothes the flanks of the hill.

Although the length of the layout from the pool to the north slope of the hill is but 140 metres, there is space for lateral extension, eastward and westward. In Tüxen's plant sociological layout in Hanover, there were about 45 associations in 1.5 hectares and ten of these were forest associations. With the lateral extensions, the woodlands which are already established in the Dundee unit would become corridors serving as links with the path by the stream which forms the central axis of the layout. Such an arrangement would clearly facilitate future management. Where, for example, a woodland in the course of development proved too small to accommodate additional associations, or not large enough to create a suitable and stable environment for the component plants, or even too small to create the illusion of genuine natural woodland, space would be available for extension while retaining the woodland in its logical altitudinal sequence along the path by the stream. But the ultimate size of these woods will still be small enough in such an open site to require their exposed margins to be closed with a dense shrub layer in order to shelter the ground layer from wind and conserve humidity within the tree canopy.

The early management of the Dundee layout has been dominated by the very high proportion of weed seeds in the soil and without modern herbicides the staff of three and the curator would have found maintenance impossible. Even with the use of herbicides, there are limits to their application in regulating the composition of the associations. For example, there are many herbicides for eliminating dicotyledons from among monocotyledons, but none which are selective against monocotyledons but leave the dicotyledons unharmed. A recent development, the 'Croptex' glove, is proving useful in singling out for treatment with herbicides individual unwanted plants growing with others in dense clumps.

A pad of fabric on the palm of the glove is wetted with a translocated herbicide, glyphosate, so that when the plant is grasped and the glove pulled along it, the herbicide is deposited. The grasping action re-charges the pad by pumping herbicide from a small container attached to the operator's belt. This method is especially valuable in eliminating plants with underground perennating organs.

The woodlands will require thinning as they develop to regulate the amount of light reaching the lower layers of vegetation, and some of the plants in these layers will require controlling in order to secure adequate representation of the constants in the associations. For example, a recent management operation was the suppression by cutting of vigorously growing *Calluna* in order to increase the representation of *Genista anglica*, which has since begun to regenerate by seeds in the small cleared patches.

The question most frequently asked by visitors is where the plants were obtained. The answer is that they were almost all threatened in the field by encroachment upon their environment. The Scots Pines were growing along the route of electricity transmission lines and would have been cut down had they not been removed to the Garden. The Oaks were rescued from a proposed road-straightening project, since abandoned! The *Goodyera* and several species of fern were dragged to the edge of a forest compartment during forestry operations and the Holly Fern, *Polystichum lonchitis*, was dislodged from a hillside and in danger of being buried by earth-moving equipment during road-widening operations. *Primula vulgaris* and *P. veris* were both introduced inadvertently with the Oak trees and both are now regenerating in the Oakwood. Many aquatic plants appeared spontaneously soon after the pools and stream were constructed. Among those found in the stream are *Callitriche hermaphroditica*, *Myosotis secunda* and *M. caespitosa*, *Ranunculus hederaceus* and *Rorippa microphylla*. In the pool, *Carex rostrata* and *Sparganium erectum* var. *microcarpum* were also introduced, possibly by water fowl which arrived at the pool as soon as thickets of *Typha*, *Glyceria* and *Phragmites* provided cover for them, a few months after being planted.

Access throughout the layout is at present provided by clearly defined pathways, the main one along the stream being surfaced with lawn grass which will ultimately become a woodland track, surfaced with forest floor litter as the neighbouring woodland develops. The other main pathway running through the layout is surfaced with *Deschampsia flexuosa*, which is not mowed but merely has the flower stems cut off before seed production to prevent colonization into the neighbouring woodland. In some places along the margin of this path, *Pyrola media* from the woodland is actually growing out into the *Deschampsia* turf. Where it is necessary to restrict access to plantings in the woodland, small accumulations of dead, lichen-encrusted branches are placed on the ground and deter by their 'cattle grid' effect. In the layout, all the circumstances of a natural environment are simulated as far as it is possible to do so in conditions of cultivation.

REFERENCES

SIMMONS, J. B., R. I. BEYER, P. E. BRANDHAM, G. LL. LUCAS & V. T. H.
PARRY (eds) (1976). *Conservation of Threatened Plants*. Plenum Press, New
York and London. xvi+336 pp.
TÜXEN, R. (1953). Etablissement et Entretien d'un Jardin Phytosociologique.
*In 'Colloque International de l'Union Internationale des Sciences Biologiques
Sur l'Organisation Scientifique des Jardins Botaniques.'* U.I.S.B. Sér. B. No.
13, Paris. Pp. 251–254.

USSR: The Policies of Botanic Gardens and their Activities in the Conservation of Threatened Plants

E. E. GOGINA*

Threatened Plants Committee of the USSR Botanic Gardens Council, Moscow

It is evident now that a settlement of contradictions between the speedily developing human society and nature has become one of the main global problems. In this connection a large system of measures aimed at conservation of nature and the plant kingdom as its leading component is being worked out in the USSR. Special articles of the new State Constitution concerning the conservation of nature are being adopted for implementation. This complicated task requires elaboration and efficient combination of various forms and methods of nature protection. Unification of thoughts and efforts of different specialists is highly desirable in this situation and the Soviet botanists are eager to do their best in this common cause.

There are 115 botanic gardens in the Soviet Union located in different parts of the country and united by a special Council, which co-ordinates their research work and promotes their mutual communication. They belong to different departments and institutions and vary in staff and size, but as a whole have a considerable number of scientists traditionally dealing with native flora and vegetation.

About 21,000 species of higher plants, belonging to widely different ecosystems—from tundra to desert—constitute the flora of the Soviet Union. The regions which are floristically the most rich—the Caucasus, Middle Asia and the Far East—have the greatest percentage of rare and endemic plant species.

The degree of disturbance in the flora and vegetation varies in different parts of the country and depends on the ecosystem as well as on the form and intensity of pressure from man and his society. Profound changes have taken place in densely populated industrial areas and in regions of intensive agriculture. The situation has become very acute in the steppe zone due to overall ploughing. More than 200 species of the native flora are in immediate danger of extinction and up to 2,000 species are in need of constant monitoring or some protective measures.

In this context, when a number of native species best adapted to the local environment are threatened with extinction, when there exists, so to speak, a flowing off of the basic botanical capital, the concern for flora conservation has become one of the main trends of the USSR botanic gardens activities. They can do much in this field and are deeply interested in it.

*Read by Mr H. Synge

141

To promote and co-ordinate their work in the conservation of threatened plants the USSR Botanic Gardens Council set up in 1974 a special committee for threatened plant conservation. Its head, Academician N. V. Tsitsin, has done much to determine principles and forms of conservation work for the USSR botanic gardens. Their work in this field has several trends. Some of them—cultivation of rare and threatened plants for conservation purposes and education of visitors—are specific for botanic gardens; others are carried out in co-operation with various institutions and organizations.

The listing of threatened plant species on a national and regional scale is one of these common tasks. Many scientists from the botanic gardens have taken part in compiling the list of native plant species to be protected in the USSR. Their data was used both in the *State Red Data Book* which is to come out this year and in the first edition of the annotated list of the threatened species of the flora of the USSR, published in 1975 under the supervision of Academician A. L. Takhtajan. The Threatened Plants Committee of the USSR Botanic Gardens Council has collected about 1500 proposals for the second edition of this book, which will also include extended regional lists.

A number of regional lists were drawn up by botanic gardens exclusively. For instance, I can mention the lists of plants to be protected in the Crimea, including 310 species, in the Altai 324 species, in the Stavropol region 171 species, in Kazakhstan 286, and in the Moscow region about 400 species.

This work has stimulated interest in local floras and more careful studies of their dynamics. Field work and expeditions conducted by botanic gardens (about 80 expeditions a year) help to find new localities of rare species and to get more information on their distributions and population sizes. The field observations also enlarge the knowledge on the ecology of rare and threatened plants, their connections with definite types of plant communities, competitive abilities and causes of decline. These observations are used in the selection of the most valuable territories for further preservation and in determining the means of protection, as well as also greatly helping to work out methods of cultivation.

In the USSR as well as in other countries protection of natural habitats is unanimously considered the best means for threatened and rare plant conservation, and areas for preservation are chosen with regard to plant distributions. Particular attention is paid to the species to be protected on the national scale, especially to the most endangered of them.

The botanists of some gardens take an active part in the selection of areas for full or partial protection. A number of these areas (including the Pripjat reserve in Belorussia which occupies 60,000 hectares) have been set up on their initiative and recommendations; background data to identify new botanical preserves in different parts of the country are being collected.

This work is not only of practical importance, for in its course some theoretical aspects of the conservation of threatened plants in nature are being developed. The concepts of complex interdependence of different plant communities in space and time are to be taken into account while singling out areas for preservation. Measures for the conservation of diversity in flora and vegetation should be related to measures for maintaining the productive functions of vegetational cover and its rôle in stabilizing and improving the quality of the environment. Dynamics of ecosystems should also be taken into consideration.

This concerns the size and the regime of the selected reserve areas and the correlation between cultivated land and wilderness.

The full diversity of any local flora reveals itself only in a complete set of the plant communities representing different endogenic and exogenic stages of plant succession. The rare and threatened species may be found at any stage of succession, though at present the last (climax and pre-climax) stages suffer most under human impact. Singling out large zones corresponding to the size of fundamental flora units with a careful policy of vegetation usage and with smaller plots more strictly protected within them seems to be the best solution to the problem.

Some larger botanic gardens have preserved plots of natural vegetation on their lands, representing different types of plant communities—from tundra on the Kola peninsula to the polydominant colchidean forest near Batumi and the Pamir mountain xerophytes in Chorogh. A good sample of European oak forest is preserved in the Main Botanic Garden in Moscow. The dynamics of all these plots is under constant study. Methods of phytocoenology are used in the conservation work of some botanic gardens, for instance in the activities of the Minsk Central Botanic Gardens of Belorussia devoted to the preservation of natural vegetation in Polessje, in the work of the Tallin Botanic Gardens on landscape protection and in the investigation of allelopathy carried out in the Central Botanic Gardens of the Ukraine in Kiev.

A thorough quantitative study of rare and threatened plants, which we consider to be very necessary, is being carried out in the Crimea by the staff of the Nikitsky Botanic Gardens. Some other botanic gardens also work in this field, but on a smaller scale, paying attention mainly to the general state of the populations of the endangered species in nature. In the most acute cases when their natural habitats are in danger of imminent destruction the plants are transferred to ecologically similar sites or into botanic gardens.

The best policy is, of course, to combine cultivation of endangered species in botanic gardens with their conservation in natural sites, but such a policy is not always possible; it depends on the circumstances in each case. For some species, as we know, cultivation has already become the only way to prevent their extinction. Cultivation of useful wild plants whose populations are over-exploited in nature can reduce the pressure on them; botanic gardens can promote such ideas, working out methods of wild plant cultivation. Well-documented samples of cultivated threatened species can serve as a reserve fund for their further repatriation into the wilderness and can be used for research work. All this gives meaning and sense to the whole trend of such work.

At the same time, in the creation of living collections for conservation purposes, the emotion and excitement of a collector eager to enrich his assemblage at all costs must give way to a precise evaluation of the real holding potential of each botanic garden. And if on botanic gardens falls the lot similar to that of Noah's Ark, we must take care that this fleet should fulfil its task with minimal losses of the precious cargo.

As a rule, the study of rare species in nature should precede their introduction. This helps to choose the best conditions for them and diminishes the percentage of losses. Seeds and living material for introduction of rare and endangered plants should be collected with the utmost care on a limited scale in order not to reduce their populations *in situ* and not to undermine their ability for self-reproduction. To ensure constant and careful observance of these principles the Threatened Plants Committee conducts the associated explanatory work.

In order to register the rare and threatened species already cultivated in the botanic gardens of the USSR, a fundamental reference book is being prepared for publication. The book will have data on the origin of samples, the year of their acquisition, the number of plants in cultivation, the completeness of their life-cycle and their reproductive ability in cultivation. The book will not only show the number of rare and threatened species already in cultivation and the activities of different botanic gardens in this field, but will also have methodological significance since alongside the analysis of the collected data will be chapters devoted to the organization of further work. The records system for conservation-orientated collections, methods of introducing rare plants and preventing their genetic changes in cultivation as well as the principles and programme for future work, will be considered.

Nowadays more than 700 species of rare and threatened plants are cultivated in the USSR. They are represented by about 3,600 samples of documented origin from different natural localities and botanical collections. Herbaceous plants prevail among them; woody plants constitute slightly more than a fifth of their number. Some larger botanic gardens have in their collections a considerable number of conserved plants (more than 200 species), whereas smaller gardens cultivate several only. About 350 species are cultivated in a single botanic garden only, while some species are replicated in many living collections; 90 species are represented in more than 10 gardens, 16 in more than 30 and 3 of them are replicated in more than 40 botanic gardens. Of the 172 species which (according to the first published list) are in the immediate danger of extinction, 96 are cultivated in botanic gardens. Among them are 5 species considered as probably extinct.

Most of the species are represented in cultivation by specimens received from the wild; only some of them were raised from the exchange of material. The data collected show that the number of cultivated rare and threatened species (and especially the number of their samples) has increased considerably in the second half of the 20th Century. The number of new acquisitions grew sharply in the 1960s and 1970s, as a result of the general policy in this field recommended by the USSR Botanic Gardens Council. However it should be mentioned that the botanic gardens had long before played the rôle of refuges for threatened plants, though on a considerably smaller scale—about 20 rare species (mostly woody ones) have been cultivated in them since the end of the 19th Century.

Some rare wild relatives of cultivated plants are also among the species being preserved in gardens. Such species and primitive varieties may have some valuable qualities lost in the process of selection and are therefore of great interest for improving the cultivated forms. They are sometimes met in the holdings of botanic gardens, but as a rule are cultivated on the experimental agricultural stations of the N. I. Vavilov All-Union Institute of Plant Industry, which is to take part in the book mentioned above.

Much attention is being given to increase the survival of the rare plants and ensure their reproduction in cultivation. The fact that botanic gardens are situated in different zones helps in the choice of the best ecological conditions for individual plants by means of exchange. Many high mountain plants, for instance, are successfully cultivated in the forest zone, whereas in the botanic gardens of their own region, situated as a rule in the lower dry parts of the mountains, they do not do well. For a better estimation of such influences on plant survival a joint experiment on growing homogeneous plant material in different zones is being carried out under the guidance of the Dendrological Committee.

In order to make the survival of rare plants more secure in some botanic gardens (in Stavropol, Donetsk, Novosibirsk), the species are grown in artificial plant communities similar in structure to the natural ones.

The work with rare and threatened plants in botanic gardens is not limited to the passive maintenance of their number or to its increase. It also aims at their comprehensive study. Cultivation enables their reproductive system to be better known and study in their ontogenesis is especially valuable as far as the rare species of almost inaccessible regions are concerned. The research carried out in botanic gardens has already provided new data on some rare species of the flora of the USSR (for instance in genera *Tulipa, Galanthus, Eremurus, Iris, Rhodiola* and *Thymus*).

Further development of this trend of work was discussed at the plenary session of the International Association of Botanic Gardens in 1975 and at sessions of the USSR botanic gardens meetings. Envisaged is an official responsibility of individual botanic gardens for cultivation of particular plant species or groups. The distribution of responsibilities will be based on the urgency of conservation measures, the scientific and practical value of the species, the means and interests of the botanic gardens and the correspondence of their climatic conditions to the ecological requirements of the selected plants. Suggestions on cultivation of different species coming to the Committee are to be discussed and approved after a proper analysis of the data on the species already provided within the cultivated reserve.

Since the work with specialized collections yields as a rule the best results, a similar trend seems advisable in the case of rare and threatened plant cultivation. Replicate cultivation, though costly, makes the conservation of species more secure. It may be justified for the most valuable or for the most endangered species. In such cases it is important to avoid accidental acquisition of alien plant material and to use the replicated cultivation for reflecting the infraspecific diversity of plants. Species introduced for the first time can substitute for the specimens widely distributed in cultivation especially for those having unreliably documented origin. It can give a considerable economy of room and labour.

Along with work aimed at conservation of local flora and vegetation special inspections of old parks and public gardens are being carried out in regions rich in them. The most valuable of them—being in good condition, picturesque and having a large range of species—are taken under protection. Such inspection of the country and town parks and gardens was completed recently in the Ukraine, the Baltic republics and in the Moscow region as well as the listing of woody introductions on the Black Sea coast of the Caucasus. In the course of this work many valuable and rare introduced plants were detected and taken under protection.

A very important trend in the activities of the botanic gardens is connected with their educational work, since the efficiency of all conservation efforts in the end depends on the degree of the public's consciousness. Information about the necessity for plant conservation is conveyed to visitors of botanic gardens on the one hand and by means of exhibitions and various forms of lecturing at meetings and in the mass media on the other. All these forms of propaganda are aimed at bringing up the public in the spirit of care for the plant kingdom. This work has a noticeable effect, but there is still much to be done on a larger scale. A steady increase in the number of visitors to botanic gardens opens up vast possibilities. Lecturing for specialists in public education—school teachers, heads of voluntary societies for nature protection and tourist associations—is highly desirable to hasten the dissemination of knowledge in the field of plant and nature protection.

An unmanaged amateur collector can do much harm to rare and disappearing plant species. Thus it is extremely important to get into contact with amateurs, constantly educate them and put their activity under the control of botanists.

Alongside education for the public at large, botanical erudition is spread in the course of contacts with state planning bodies and local authorities. In many regions, following recommendations by botanic gardens, rare and threatened plants of local floras have been taken under protection, their storage and sale have been prohibited and their locations as well as the most valuable old parks and gardens in which they occur have been declared to be reserves or natural monuments.

Proceeding from the assumption that joint action of all nature protection organizations and institutions largely determines success, the USSR botanic gardens develop co-operation in this field with a great number of them. They share experience at their annual sessions, during their joint expeditions, in the press and in various other forms. The plenary session of the International Association of Botanic Gardens held in Moscow in 1975 was devoted to the problems of plant conservation. It addressed all botanic gardens of the world with an appeal to co-ordinate their efforts and develop co-operation in this field. It was met with a warm response among the Soviet botanists and they have the sincere intention to do their best to strengthen international contacts in the field of plant conservation. With this aim Soviet botanists visited Czechoslovakia and Poland and received foreign guests in the USSR. At the scientific conferences of the member countries of the Council for Mutual Economic Assistance (CMEA), botanists exchange information on plant conservation. For three years now joint Soviet–American botanic expeditions have been carried out. During these expeditions the field status of rare and endangered plants and the causes for their decline are studied, materials for introduction collected and experience on their conservation exchanged.

The importance of world-wide exchange of the experience is beyond doubt and I am sure that this needful Conference will do much to help botanic gardens of different countries to fulfil their duty in the crucial task of flora conservation.

REFERENCES

BORODIN, A. M. *et al.* (eds) (1978). *Red Data Book of USSR*. Lesnaya Promyshlennost, Moscow. 459 pp. (In Russian).

TAKHTAJAN, A. L. (ed.) (1975). *Red Book: Native Plant Species to be Protected in the USSR*. Leningrad. 204 pp. (In Russian).

USSR: Rare and Threatened Plants and their Conservation in the Botanic Gardens of Kazakhstan

B. A. WINTERHOLLER

Central Botanical Garden of the Kazakh Academy of Sciences, Alma-Ata, USSR.

The plant kingdom of Kazakhstan is rich and diverse. The local flora numbers 5630 higher plant species, among which are 70 tree species, 700 species of shrubs, dwarf shrubs and semi-shrubs, and over 4500 herbaceous species. Such richness of the flora can be explained by the surprising diversity of natural conditions due to the complex relief and different climatic conditions of the Republic.

Plant cover of Kazakhstan takes the form of different types of forest, steppe and desert. Among them are not only unique but also relict formations which need more thorough conservation. These are some types of *Stipa* steppes, riparian forests (especially Asiatic poplar forests), ash forests, pine forests on the southern edge of their area (Kzylrai), mountain wild fruit forests in the south of Kazakhstan, the most complex types of sandy vegetation, savanna woods, light forests and scrub of Karatau, Betpak-Dala and others. At present many of these are suffering severe damage.

The flora of the southern and eastern mountain regions of Kazakhstan, where there are many rare, endemic and relict plants, is very diverse. Kazakhstan has one of the highest numbers of endemic species in the Soviet Union with over 800 endemic species; these plants are of particular value for science. Endemism in the flora is not only shown at species level but also at generic level. For example, in Kazakhstan there are such endemic genera as *Spiraeanthus, Stephanocaryum, Pseuderemostachys, Ikonnikovia, Pseudomarrubium, Ugamia* and others.

The conservation for the future of ancient relict species, which are vivid witnesses of evolution, is extremely important. A series of such plants is of particular interest as the gene pool that will be used sooner or later. Among these species are *Spiraeanthus schrenkianus, Incarvillea semiretschenskia, Ikonnikovia kaufmanniana, Ostrowskia magnifica, Atraphaxis muschketowii, Atraphaxis teretifolia, Cancriniella krascheninnikovii, Pseuderemostachys severtzovii, Pseudomarrubium eremostachydioides, Zygophyllum subtrijugum, Pastinacopsis glacialis, Stroganowia trautvetteri, Physandra halimocnemis* and others.

Natural stands of relict xerophilous light forests and shrub steppes remain only in the mountains of Syrdarinsk Karatau and in the central and western parts of the Betpak-Dala desert. Their light stand communities much resemble savanna. These shrubs are well adapted to life in desert conditions, having no summer dormancy, and vigorous root systems which reach ground water. In this connection *Spiraeanthus schrenkianus* is of great interest as a prospective plant for desert afforestation. The most ancient relict representative of the Kazakhstan flora is *Incarvillea semiretschenskia*, which can be seen only in the Chu-Ili mountains. This unique plant was first found in the Semiretschensk district in 1905 by the expedition of V.E. Niedzwedzkii, the well known explorer of Middle Asia and Kazakhstan. In 1915 the botanist B. A. Fedschenko provided the botanical description of this plant and distinguished it as the new genus *Niedzwedzkia*, with the single species.

Among the wild flora of Kazakhstan there are plants which are relatives of cereals, e.g. *Hordeum bulbosum, Hordeum spontaneum, Aegilops cylindrica, Secale silvestre*; there are also many fruit-bearing plants such as *Malus sieversii, Malus kirghisorum, Malus niedzwedzkyana, Armeniaca vulgaris, Vitis bosturgaiensis, Allium pskemense, Allium longicuspis*, which yearly produce thousands of tons of valuable food products. Medicinal plants of the Republic number over 500 species and out of them 50 species are used for the preparation of medicinal raw materials. These 50 plants include *Artemisia cina, Allochrusa gypsophiloides, Ephedra equisetina, Glycyrrhiza glabra, G. uralensis, Rhaponticum chartamoides, Rhodiola rosea*, and the number increases every year. For example, *Artemisia cina*, a highly effective medicinal plant that contains 6.5–29 per cent of santonin, has been used for about 100 years, whilst soaproot, *Allochrusa gypsophiloides*, that contains about 30 per cent of saponin, since 1927. Further industrial usage of these plants may bring them to the ranks of the disappearing ones and the exploitation of natural stands of medicinal plants leads inevitably to impoverishment of their gene pool. That is why the most valuable and rare species that have narrow distributions must be introduced into industrial cultivation.

The significant transformation of Kazakhstan that has taken place over the last few years has greatly influenced the distribution of many plants. Intensive reclamation of the territory of Kazakhstan over the last few decades, such as ploughing of virgin lands, building of new industrial centres, increasing the transport network, are the most powerful factors that have brought about changes in the flora.

At present under direct or indirect human influence many species of the local flora of Kazakhstan have become rare or are disappearing, as for example are beautiful early-flowering ornamental tulips, crocuses and irises. Mass picking of wild flowers causes fatal damage to plants in the wild. In this way *Tulipa ostrowskiana, T. kolpakowskianum, Crocus alatavicus, Iridodictyum (Iris) kolpakowskianum, Paeonia hybrida* have become rare and have disappeared from the suburbs of Alma-Ata. A decrease in the stock of ornamental plants is very noticeable in places of rest, in the suburbs of large cities and industrial centres, as in for example Karaganda, Ust-Kamenogorsk, Leninogorsk, Djambul and Chimkent.

Taking into consideration the importance of measures to conserve the gene pool of Kazakhstan vegetation, botanic gardens of the Republic fulfil their rôle by working on distinguishing floristic composition, studying the ecology and distribution of rare and threatened plants of the local flora and their introduction. In recent years several floristic regions rich in rare and endemic plants such as the Chu-Ili mountains, the Karatau, the Zailiisk and Talas Alatau, the valley of the Ili river have been explored. The natural habitats,

ecological conditions and distributions of rare and threatened plants of Kazakhstan are being ascertained. Large-scale mapping of standard plots is made. The
general state of populations and the biological peculiarities of rare species are
described.

The displays in the systematic collection of the native flora in the Central
Botanical Garden of the Kazakh Academy of Sciences contain more than 70
species of rare and endemic plants. *Incarvillea semiretschenskia, Amygdalus
ledebouriana, Aflatunia ulmifolia, Atraphaxis muschketovii, Crataegus almaatensis, Morina kokanica, Ikonnikovia kaufmanniana, Korolkowia severtzowii,
Berberis iliensis, Iridodictyum kolpakowskianum, Iris albertii, Crocus alatavicus,
C. korolkovii, Tulipa greigii, T. kaufmanniana, Juno (Iris) almaatensis* and *J.
kuschakewiczii* have been successfully introduced and are recommended for
cultivation as ornamental species.

Valuable medicinal and edible plants of the native flora, now becoming rare,
like *Allochrusa gypsophiloides, Artemisia cina, Artemisia transiliensis, Allium
longicuspis, Allium altaicum, Astragalus glycyphyllus, Malus sieversii*, have been
tried in cultivation. Among the rare and endemic species of central Kazakhstan,
*Berberis karkaralensis, Caragana bongardiana, Iris lactea, Linum pallescens,
Betula karagandensis, Papaver tenellum, Allium inops, Caragana balchaschensis,
Tanacetum ulutavicum, Betula kirghisorum*, have been studied in the botanic
gardens of Karaganda and Dzhezkazgan. A collection unique for Kazakhstan
of 18 fern species deserving protection has been created in the Altai Botanic
Garden; these include *Dryopteris cristata, Athyrium rubripes, Polystichum
braunii, Cryptogramma stelleri, Polystichum lonchitis* and *Camptosorus sibiricus*.
The biology of rare and endemic plants of eastern Kazakhstan such as *Tulipa
altaica, Allium platyspatum, Erythronium sibiricum, Iris ludwigii, Pyrethrum
kelleri* and *Orobus latyroides* has been investigated.

The Central Botanical Garden renders great assistance to societies for nature
protection in detecting unique natural areas and in organizing preserves. The
list of endangered native Kazakh species for further state protection has been
worked out and consists of 286 species. This information is used in compiling
the *Red Data Book* of the Kazakh Soviet Socialist Republic. An exhibition of
Kazakh rare and disappearing plants, presenting about 200 species of rare,
endemic and relict native plants, is being prepared in the Botanical Garden.

Kazakh botanists take part in all-union and republican conferences, symposia
and meetings devoted to the problem of nature protection and rational use of
natural resources, as well as in all-union seminars where experience in the
conservation of rare and medicinal plants of the USSR is shared. They also
participate in the exhibit on 'Rare and disappearing species of the USSR and
their protection' at the Exhibition for Economic Achievements.

The results of the research in the field of rare plant protection are popularized
in the scientific literature, and through articles and posters, and by lecturing on
radio and television. Considering the great economic importance of rare and
disappearing plants, the Kazakh botanic gardens are planning to become more
active and broaden their scientific research work in plant conservation.

USSR: Rare and Protected Decorative Plants of the Wild Flora of Georgia and their Cultivation in the Tbilisi Botanical Garden

L. V. Asieshvili

Tbilisi Central Botanical Garden, Academy of Sciences of the Georgian S.S.R.

Great variability in soil and climatic conditions together with the natural terrain of Georgia have resulted in the evolution and survival of a great number of plant species, amongst them many endemics of a wholly local habitat. According to Academician A. A. Grossheim (1940), the wild herbaceous flora of the Caucasus consists of 1300 species and in Georgia alone there are more than 1000 species, of which 400 are endemics. This extraordinary rich decorative flora of Georgia has for a long time attracted the attention of botanists and other scientists from different European countries.

During the last century there were many beautiful flowering plants of the natural flora of Georgia that were brought into cultivation in foreign countries, where many of them such as *Delphinium, Primula, Paeonia, Tulipa, Lilium* and *Scabiosa* are still grown as commercial crops. As well as direct introduction into cultivation many of the decorative plants of Georgia have, due to the high quality of the initial material, been used in the selection and hybridization of new cultivars. It is therefore, in our opinion, only natural that a great deal of attention should be given to the conservation of the natural flora particularly to the rare and diminishing species, which can become a priceless gene pool for scientific and horticultural research.

Conservation and preservation of the plant kingdom is an important component of environmental protection and has become State Policy in the Soviet Union. In this serious pursuit the leading rôle is taken by the scientific institutions, which must take responsibility for restoring the vegetation, finding out which plants are rare or in danger of becoming extinct and introducing them into cultivation, and also in finding the best ways of using plants in both the scientific and economic activities of man. Botanic gardens in the Soviet Union are taking a lion's share of responsibility in this task. In these gardens there ought to be present those species that are rare and those threatened with extinction; also these gardens should draft scientifically based measures that lead to the preservation, reproduction and prevention from total extinction of all such species in nature. This aim is well reflected in our endeavours at the Central Botanical Garden of the Academy of Sciences of the Georgian Soviet Republic, where there is a special area of land within the living collection devoted to rare, mostly herbaceous species of the flora of Georgia and the Caucasus, bearing a symbolic name 'The Kazbeks'. Here a variety of ecological niches is provided for growing

FIGURE 1. A general view of the 'Kazbeks' site.

FIGURE 2. *Primula juliae*

FIGURE 3. *Asphodeline taurica*

FIGURE 4. *Veronica telephiifolia*

plants collected from various habitats. The experimental 'Kazbeks' site is situated on the periphery of the Botanical Garden under a rock outcrop, almost 560 m above sea-level. The physical relief of the site resembles the natural landscape of the Great Ridge of the Kazbek Mountain with its canyons, precipices, high slopes, etc.

The selection of plants to be grown is made with regard to their economic and decorative importance, spreading pattern within the habitat, abundance on sites

FIGURE 5. *Paeonia lagodechiana*

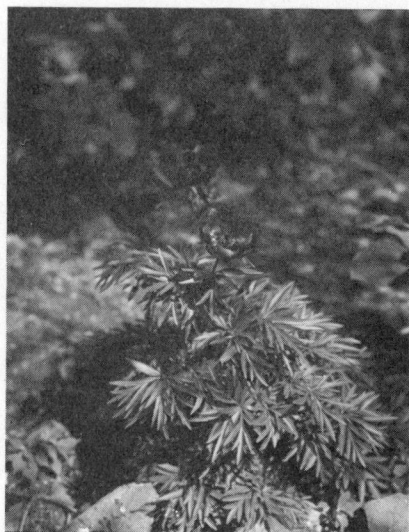

FIGURE 6. *Paeonia majko*

where they are thriving, and ability for natural regeneration. Special attention is given to local endemics, which are of great scientific value.

Many years studying our flora has enabled us to pick out roughly 400 plant species which ought to be included in a supplement to the *Red Book*. Many of these species, brought in as living plants from the various mountainous regions of Georgia and the Caucasus or grown from seeds, are thriving under cultivation at the 'Kazbeks' site.

Among those species which are decorative, which are rare and which have been comprehensively studied in cultivation over many years on the 'Kazbeks' site and requiring protection, we would recommend: *Paeonia lagodechiana, P. macrophylla, P. majko, P. mlokosewitschii, P. wittmanniana, Primula juliae, P. woronowii, Campanula mirabilis, C. kemulariae, C. kolenatiana, C. raddeana, C. sarmatica, C. ossetica, C. petrophila, Cladochaeta candidissima, Salvia garedji, S. daghestanica, Iris iberica, Eremurus spectabilis, Asphodeline taurica, A. lutea, Dianthus imereticus, Pyrethrum sericeum, Scabiosa caucasica, Anthemis saguramica, Lilium ledebourii, L. monadelphum, L. szovitzianum, L. georgicum, Papaver bracteatum, Muscari szovitzianum, Tulipa eichleri, T. biebersteiniana, Veronica telephiifolia, V. armena, Gypsophila tenuifolia, Sempervivum pumilum, Scorzonera ketzkhovelii, Omphalodes cappadocica, Symphyandra pendula* var. *transcaucasica* as well as many others.

There are more than 350 species of plants gathered on the site, all subject to protection. This figure gets larger and larger year after year as new rare species are brought into cultivation.

PART THREE

Education

Public Appreciation: Who Digs Botanic Gardens?

DON ALDRIDGE

Countryside Commission for Scotland, Perth

This paper considers the potential which exists in botanic gardens for interpretation to the general public, in particular to the family visitor.

Since European Conservation Year in 1970 many managers of botanic gardens, museums, wildlife parks and zoological gardens have made a reappraisal of their objectives. In 1976 the Council of Europe published *Environmental Awareness*, a survey of such facilities in the member countries, which looked at the possibilities that exist for co-operation between facility providers. (For example, the Royal Botanic Garden at Edinburgh, the Royal Scottish Museum and the Royal Zoological Society of Scotland work together on a project called 'Interlink', and Zoo-Skoletjenesten in Copenhagen operates a co-operative education service with the Lejre Archaeological Research Centre, the Viking Ship Museum, and the museums of the City.)

Whilst it is important to give pride of place to the traditional objectives of botanic gardens it is worth looking afresh at what is offered to the public. The range of possibilities varies from the well known interpretive work of the United States National Park Service to the equally famous Arizona-Sonora Desert Museum, whose stated aims are listed on the following pages.

Of course there are many excellent European facilities which could not be mentioned in the Council's Report. Some, like the Hortus Botanicus in Amsterdam, seem so limited in size and scope for future development that its directors might be expected to give up trying to gain visitor attention and interest. Not so! This very small botanic garden must rank as one of the most interesting in the world on account of its exhibitions, its guide book and the thought which has gone into its use of presentation media. Success has been achieved without loss of authenticity, without introducing gimmicks, and without a proliferation of gadgetry. The communicators understand that it is the plants that are of interest and are important!

The following pages are an abbreviated version of an audio-visual programme given at the conference.

157

Botanic Gardens by Don Aldridge

Respected and highly expert audience of famous botanists! Why have I been invited to speak to you? One can only guess.

1. I am uniquely ignorant of all things botanic but what is worse I have already passed on to my offspring an IGNORANCE GENE

2. I am rare, unique even

3. Sometimes I show signs of being an incurable romantic (though not a thing of beauty) with just the merest hint of exclusiveness

4. and now I am at the limit of my range. I apologise for drawing in the middle of your serious seminar.

RARE

5. and one other qualification is that I am endangered and only very recently I was threatened!

6. Lastly I am finite!

......messages?

So those are my SIX messages which I think you might communicate to the public — the "why" of conservation — the philosophies which you could make practical:

a. the idea of keeping options open (for future choices) or the idea of the gene reservoir
b. the idea of rarity and loss to the scientific record
c. the idea of loss to the heritage (a mish/mash of educational, social, scientific & romantic elements)

d. the idea of variety (well I suppose there are some ecological principles) and coupled with this
e. the general idea of diversity which should not be threatened
f. and likewise the paradigm of the vital relationship between population and resources

Actually I do have a garden. People laugh at me but it is botanical. It has no weeds and no pests — because these are unwanted things.

But a near neighbour is a high priest of the British religion of gardening.

He does everything that the weekend magazines say you should do in your spare time. But his evangelism and crusades have no effect on me as I have no spare time and opening botany books puts me off and Botanical Gardens actually frighten me.

Sometimes the didacticism actually drives me away especially when they are at pains to tell me what things are NOT!
And figs are not fruit and bamboo is not a species, salaginella are not mosses, cyads aren't ferns, mosses aren't lichens, milkweed is not a cactus!!!

160

But as you are already into adult education maybe you could
win more attention if you took one of the three subjects
our schools are supposed to avoid. Here are some
examples which I have seen.

Politics

As one European Botanical
Garden simply states:
"tea picking is always done
by women"

Religion

"Plants" says the label in another garden "were put here for man's use"

og! Number 4711 is toxic!

So much for teleology mum!

Sex

Another European garden reveals
the secrets of the sex life of
Stenocarpus salignus. . . the bottlebrush
which is pollinated by Male Honeymice
in the middle of the night!

yes it's
true!

News
of the
World
scandalous!
says Bishop
in
eye-witness
account

It is no laughing matter!
Wallace's observations of this insect led
him to predict the existence of the Deutsche Gramophon, this fly reaches the parts that others cannot.
So can you!

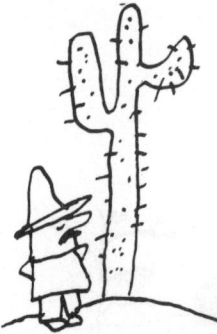

Consider, for example, the splendid objectives of
the ARIZONA — SONORA Desert Museum:
   to enable visitors to gain knowledge in a few hours
   that would otherwise require years of wilderness
   searching

   to foster appreciation and knowledge of living
   things

   to awaken public interest in resource conservation

   to stimulate and aid scientific research in
   natural history and resource conservation

   to operate a centre for outdoor education
   and research

Consider also the interpretation work of the United States
National Park service with its emphasis on

| The message repeater | the interpretive sign | the earplug soundguide | and the radio listening post |

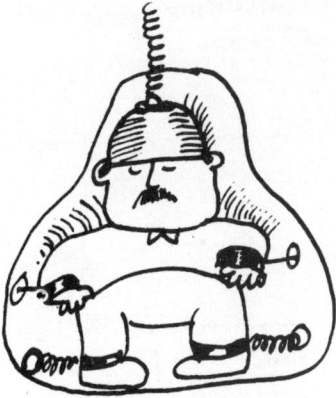

and the ultimate in media,
the listening chair
but you could invent

new media,
new botanical
message repeaters,
books even

Papyrusiet

PLEASE
do not feed
the
PIRANHA
fish

← leave a small pair
of boots and socks
beside the notice

But new media need not be an invitation to vandalism. You must
word your notices carefully.
Apart from media don't forget you have another resource: Staff, old & young

That is Higgins
He has been with us
since the Carboniferous

163

been at the hallucinogens again have we lad?

Please don't ask the staff questions — THEY MAY BITE WHEN PROVOKED

So perhaps some Botanical Gardens will be tempted to interpret

The Moses Creche
why not leave your infants in the bullrushes?

To the GARDEN of EDEN
Adults only

or follow the Council of Europe's "Facilities Approach" and co-operate with Zoological Gardens but let us hope they won't forget
their main objectives in the process!

The Work of the Hanover School Biology Centre as an Example for Teaching Methods in Botanic Gardens

GERHARD WINKEL

School Biology Centre, Hanover, Federal Republic of Germany

INTRODUCTION

An institution can fulfil its current tasks only in the light of its history. This is a characteristic of all living systems. Although it is one of the basic insights of ecology, it is occasionally overlooked today in connection with social reforms. Our model from Hanover can be adopted only in its basic principles, and not in its totality.

The motivation for our work is environmental education, in the belief that an understanding of the results of science, combined with the emotional experiences gained from studying live plants and animals, leave a lasting impression on pupils, and thus our aim is teaching with living nature.

The tradition of the School Biology Centre dates back to 1883, when the institution was founded at the suggestion of teachers, to supply the 33 schools then in Hanover with equipment for botany. The ups and downs of its history cannot be described here, but have parallels in many botanical institutions. New life was breathed into the institution in 1961.

The Centre currently consists of the School Botany Garden in Burg (7 hectares), the School Botany Garden in Linden (c. 2 hectares) and a Zoo School attached to the Hanover Zoo. There are about 1000 sq. m of glasshouse space as well as a set of classrooms and a collection of educational material. The staff total 31, which includes five teachers and 19 gardeners. The Centre is under the school administration and the City pays salaries and wages. This enables the Centre to provide about 130 schools with the services described here.

SUPPLY SERVICE

In January the schools receive schedules of what can be supplied at elementary and secondary levels. Each schedule covers 15 themes, with all the details, and the list of materials accompanying each subject gives an evaluation of their instructional value.

Teachers may order a complete package for a particular theme or make up their own. At elementary level, we offer material for studies on genetics, as well as on morphology and seed germination, while at secondary level, subjects include nutrient deficiency symptoms and evolution. For example, seeking a theme on genetics for elementary level which could be used in a variety of other ways, we have developed the comparison of wild and cultivated alpine violets (*Viola* spp.). We try to develop in the schools the specialized knowledge required for each project, along with the educational insights that each project can give. The instructional aid for the teacher is in the form of a pamphlet, which includes any appropriate didactic comments (e.g. on age level, learning goals) and details on the biology itself (including hints on how to make use of the material, showing how the various tasks may be assigned and how the literature and audio-visual equipment should be used). Charts and graphs are included to help instruction. Each pamphlet is 10 to 25 pages long and is received by the teacher in advance of the materials themselves, which are delivered by municipal transport. Sets of equipment, each consisting of as many as 20 different items, are sent to between 50 and 180 schools; this material is not taken back, but remains in the schools.

LOAN SERVICE

In addition to the supply of non-returnable material on specific themes, there is a service under the control of a teacher where schools may borrow specific items such as complete aquaria, insectaria, formicaria, measuring instruments, sets of microscope slides, etc. (The full list of equipment available is 25 pages long!) As with the supply service, all loans are accompanied by advice to the teacher on how to use the material. It has become apparent that the loan service is an indispensable help in biology teaching and, even with the shrinking resources available from our budget, provides an optimal teaching return on the investment.

VALUE OF THE SUPPLY AND LOAN SERVICES

We have found for both the loan and the supply service that:

1. Plants must be in pots, or the material must be resistant to damage.
2. All plants delivered should be related to the various themes.
3. Instructional aids should accompany the deliveries.
4. Deliveries must be punctual and conform to precise planning.
5. Liaison with the Centre should be the responsibility of one teacher in each school.

We have found that materials on loan have helped to make teaching more individual. Providing advice on how to use our facilities encourages continuing discussion, which helps the learning process. We have also found that a central loan service greatly reduces the costs of teaching natural sciences.

What relevance can this have for botanic gardens? It means that supplying plant material to illustrate particular educational themes could be carried out in conjunction with every botanic garden at a relatively low cost; the key factors are precise planning and delivery, effective advertising at the beginning, and a dedicated team. Every botanic garden could prepare various materials for loan, but a liaison officer must be hired or trained to carry out the work.

Another large field that is still neglected is research into the use of plant material in botany teaching. This should include finding the plants that are most suitable for each theme, in particular for the study of flower ecology, genetics, succulent morphology and parasite biology. Other needs are for plants with easily visible chromosomes. In all cases, the plants should be easy to grow and maintain. In the Centre we have begun to compile a dictionary of botanical concepts particularly relevant to school biology and it is worth considering whether this could be developed into the first co-operative educational effort among all botanic gardens.

SCHOOL GARDENS

There are many ways of looking after plants, a subject that is perhaps the most important basis for education on environmental protection. The gardens in Hanover's schools have had a shadowy existence for the last generation and so it is particularly important to help those teachers who are willing to take over the task of their cultivation and maintenance.

All schools receive a catalogue of available plants, arranged according to educational precepts. It covers plants of use for general cultivation, for genetics, for cut flowers, etc. Teachers receive the material they request either as seed or as small plants, depending on whether a heated frame is needed to start them off, and already 45-50 school gardens are provided with plants each year, together with general advice.

Some schools wish to plant shrubs and bushes. Therefore, in September of each year, there is an 'open week' for schools at the Centre, in which the gardener in charge shows the teachers around the cultivated plants, gives advice on their cultivation and, working closely with the teachers, produces a planting scheme for each school. A few days later, the schools receive the plants free of charge. In addition the schools can order bedding plants such as pansies and wallflowers without visiting the Centre and some 15,000 pansies are grown each year for this purpose alone.

As a result of this work we have realized that gardening in schools is dependent on the enthusiasm of individuals; it is especially important that those teachers who do look after school gardens should keep in touch with each other. We have found in particular that there is a large discrepancy between accepting the importance of these gardens, and the willingness to put the necessary work into practice. School gardens and grounds in general have not been considered so far from the truly educational viewpoint.

How much of this work could be undertaken by botanic gardens? First, activities for amateurs could be offered, e.g. courses in cultivating orchids or succulents, or in floral art. An information centre giving help on plant diseases and advice on cultivation could also be started, whilst plants for botany teaching could be given to the general public as well as to teachers and possibly school classes. In addition, botanic gardens could open themselves up more than at present by allowing amateurs to work as hobby-gardeners.

The educational rôle of gardening itself provides a large field for research and is almost an untouched continent. A few questions may be mentioned here:

1. What impact does dealing with something alive have on the development of people?
2. To what extent can gardening serve as a therapy?
3. What are the criteria for a basic course in gardening for students or laymen?
4. Why has there been a decrease of interest in cultivating plants and looking after them?

PUPIL AND STUDENT COURSES

While the basis of the delivery services is the plants themselves, the instructional ability and the personality of the teachers are most important in the courses. We arrange 90 minute tours as well as one-day and two-day courses for school classes; these cover bee-keeping (one or two days), gardening and flower arranging (two days), as well as two-week practical courses.

For simplification, only the two-day course is described here. The requests for places greatly exceed our limited capacity; 600 of the 1800 school classes in Hanover have asked for instruction but unfortunately we can only receive 150. For each course, the class teacher, the biology teacher and our teacher must first meet and discuss the course; after a guided walk around the Centre's garden, special requests are fitted in with what is available. The greatest demand is for courses on experimental behaviour, genetics, limnology, tropical rain forests and introductory microscopy. The programmes are devised for children in grades 5 to 13.

The instruction itself is done by our teachers and is object-orientated. Because of this many teachers from the schools get new ideas and very often young teachers appreciate watching another teacher instructing their class. The facilities of the Centre are used in conjunction with the course; these include animal preserves, a large model bee-hive, and garden areas in which children can study practical genetics, etc. The materials available would be adequate for studies at university level.

As a result of our course work several generalizations can be made:

1. Even biology teachers constantly need new ideas.
2. Instruction by a teacher other than the class teacher creates a strong social impetus in the class.
3. The number of teachers who want to take over instruction at the Centre is extremely low. A teacher familiar with the Centre is the basis for the success of our courses.
4. The courses have a strong publicity effect for biology as a subject and many teachers of other subjects register for instruction.
5. Since the instruction is by demonstration and experiment, the teachers tend to be successful in putting across new subjects such as limnology, ethology and plant sociology.

6. Each topic can be adapted for every age group and on every school level. For instance mentally or physically handicapped children can be taught to use microscopes expertly, and high school children can become interested in binding and arranging flowers.

7. The constant changing of classes necessitates a time in the winter when there is no teaching, so that the teacher can prepare new subjects for study.

Again how much of this could be taken over by botanic gardens? A full-time teacher would be essential and he or she must stay at the institution for at least two or three years to do fruitful work in this field. During the 200 school days every year, a teacher can give about 50–60 two-day courses, 20–30 one-day courses and 30 tours. On a realistic assumption that 20 per cent of the classes in the area would be interested, one teacher would be needed for every 100,000 inhabitants (c. 350 classes). At this rate, about 3000 students would participate each year.

Gardeners too can be used for teaching, but a specific classroom must be provided. Discussion and dialogue between gardeners, scientists and educational experts is of value to all, and the teacher must not be isolated from the practical experience of others. Pragmatically, confining the courses to botany is not justified, and it is a natural progression to expand into zoology, for instance by studying birds, animal pests, aquatic life, etc.

The following questions are starting points for research into the educational rôle of gardens:

1. What characterizes gardens that are educationally successful?

2. How do they activate their visitors?

3. How realistic and practical should the approach be?

4. What is the teachers' rôle in motivating and developing young interests?

GUIDED WALKS

We have offered a Sunday morning programme for the last three years where parents can walk through the garden at the Centre with their children. For educational reasons we insist the parents accompany their children. Every Sunday six to ten teachers are available to guide walks in the garden on particular themes; as well as the talk itself there is always the opportunity for parents to discuss matters with the teachers. Up to 400 visitors now take advantage of this every Sunday and we hope it will help to deepen the dialogue within each family about biological and ecological questions and will help to influence people's opinions and appreciation of environmental problems.

TEACHER TRAINING

The paper would be incomplete without some mention of the importance of an institution like ours in the 'coining stage' of teachers. Every teacher in the Centre is in charge of a group of young teachers as their Biology Seminar Director. Every biology teacher (except for high school teachers) is trained in the Centre for two to three years; they can take advantage of the equipment and the teaching concepts already formulated, and they can work out new concepts for themselves; in the programme there are about 70 teachers connected to the Centre. We also have one course each year for the training of school assistants.

Advanced teacher training has been neglected so far. Lack of time and a certain apathy towards advanced studies are responsible for this. However, both in Hanover itself and in Berlin there are educational centres where more extensive projects are worked out. For example, a Biology Curriculum for grades one to ten has been elaborated, problems of human ethology have been discussed and educational films are screened. Moreover, all Biology Seminar Directors from Lower Saxony meet at our Centre for one week every year and two or three groups of university students or teachers also receive training over three to four days. We also greatly support the work of the forest youth hostels in Lower Saxony. This is all that can be done; there is no more time, space or energy!

The Appreciation of the 'Conservation Rôle' by Staff of Botanic Gardens

R. I. BEYER

Royal Botanic Gardens, Kew, England

The implications arising from the decision by senior managers of botanic gardens, and allied institutions which hold living plant collections, to cultivate plants of particular conservation importance are the basis of current discussion. The purpose of this paper is to draw attention to the need to consider aspects of training that should be provided within such establishments if any measure of success is to be achieved. Many plants would become either lost or put at considerable risk without such training of the staff who in the course of their daily work are required to maintain this material. Plants grown in artificial, man-controlled environments are by definition exposed to a higher degree of risk than those growing in natural ecosystems, and it is pertinent to observe that it is almost impossible to guarantee their survival in any one garden for any length of time unless human error is reduced to a minimum.

Managers of living collections however can, and will, reduce the risk of loss considerably if the collections are recognized as of extreme importance by all members of staff from the skilled to the unskilled. The formulation and introduction of handling systems at each stage of a plant's progress from its reception to its eventual establishment are essential; but at the same time every gardener must understand the importance of such systems and apply them meticulously. A full appreciation will only be achieved by the training of new staff and by motivating longer-serving gardeners to seek a greater understanding of their particular speciality.

The organization of staff within each garden will differ in detail but the levels of skill and responsibility will be roughly the same. Each skill level requires specific tuition and training within broad bandings according to individual responsibility. The size of the collection grown and the number of staff employed also dictate the training system that should be adopted, but the one common denominator will be that each member of staff will be required to know that the plants they are helping to maintain are of considerable scientific importance and cannot be replaced readily.

NEW EMPLOYEES

The pressures created by finding oneself in unfamiliar surroundings are known to us all and I feel sure that we can recall instances which have affected our composure at some time or other. Unfortunately these pressures are often forgotten once an individual has become adjusted to new circumstances and this leads to the neglect of others who later have to suffer the same experience. Managers at all levels within any organization should, if they are to achieve a fully integrated and informed work force, take this into account by providing opportunities for new staff to be introduced and briefed with essential information at the earliest opportunity; and by doing this the initial pressures and settling-in process can be reduced to a minimum and feelings of isolation of a newcomer quickly dispelled. Failure on the part of a manager to recognize this need will inevitably lead to problems which in the long term can cost more to overcome than the initial time spent on introduction and instruction. New members of staff left to find out for themselves will be forced to copy or ask for guidance from those with whom they can most easily identify, namely their immediate colleagues. If that colleague in turn had not received adequate briefing the information which will be passed on will not be necessarily either informed or accurate.

I am of the firm opinion, which has been formed by personal experience both as a newcomer and later as a manager, that time spent in this way is regained and increased at a later stage. How often has a manager been faced with the answer "Nobody told me" when a problem has arisen? An excuse which in my experience is often justified.

BASIC SKILLS

The difficulties over the recruitment of qualified and skilled gardening staff seem to be an international problem, which perhaps is now becoming even more acute in the industrialized countries. At Kew we have great difficulty in recruiting staff who have been trained in basic skills and consequently have to employ many unskilled and inexperienced people. To compensate for this lack of skill we endeavour, not always with success, to teach basic skills from the outset by on-the-job instruction and by integrated training programmes. At this stage it is essential that the value of plant collections with which they are working is fully explained and understood. The quality of work produced and attention to detail should be emphasized and firm standards established so that the need for close supervision is reduced quickly. By exposing a new worker to initial instruction in a reasonably broad band of practices the individual's potential can be quickly assessed. The responsibility for this training is placed in the hands of the immediate supervisor so that the needs of the job will be met and so that the supervisor will know the strength and weaknesses of each individual from the outset.

SPECIFIC SKILLS

Having recognized the most appropriate job within the organization for an individual, the next stage is to ensure that consideration is then given to providing instruction in specific areas of work on which he or she is engaged. It is very easy to neglect the need for formal instruction at this stage, and we have found that there are certain common areas of work within each section which can be improved if time is spent on further instruction. To assume that because individuals have shown initial expertise at a basic level they are able to carry out more demanding tasks is, in my opinion, foolhardy. The self-taught gardener will only see the job from his point of view and if allowed to drift into bad habits will be very difficult to control. Standards must be maintained and this will not be possible unless specific instruction is given.

We have abandoned the practice of trying to train on a broad base at this stage, and have taken the attitude that it is far more productive to concentrate on teaching skills related to the tasks in which the individual is engaged. It is now an exception to find the inexperienced gardener who has the necessary qualities to become competent on a broad front but, if greater potential than normal is recognized, opportunities can be made available for development at the appropriate rate.

INFORMATION AND AWARENESS

Staff, if asked directly why living plant collections are being grown, are likely to give a variety of reasons according to their involvement in policy-making decisions. At the lowest end of the scale staff often have little idea and those that do make comment are usually badly informed. The most important point must be to explain simply about the rarity of certain species, the need to conserve them, and point out what measures will be taken to achieve this aim. When an individual has been given responsibility for a specific group of plants he should be encouraged to develop an interest in their cultivation and given every opportunity of meeting specialists in his field. At the Supervisor/Manager level a reluctance to share rarities with others (the collector's syndrome) can be a major hazard and something to be discouraged at all costs.

I must admit I have no easy recipe to create awareness other than to suggest that communication in all its varied forms is essential. Information must be made freely available but, above all, there must be an appreciation of procedures relating to the maintenance of material of conservation merit. In this context information can be made available in written form but we all know how easy it is to file paper away and forget. We have found that within our organization the first line supervisor is the key person who must set high standards of cultivation, insist that procedures are followed and ensure the staff are well briefed. For this reason it is well worth investing time and money in the training of senior horticultural staff including if possible collecting plant material in the field. Plants grown in artificial environments, in my experience, often fail because cultivation techniques are wrong and do not take into account the needs of the individual plant. Field data should include information to help the cultivator, but it must also be made available to him, and he must use it.

DEVELOPMENT TRAINING

Excursions into the field leave me with a feeling of inadequacy brought on by realizing that, after 30 years in horticulture, my knowledge is abysmally low.

As mimics of nature we create very poor artificial environments in which to grow plants and I think I can safely say we cannot create a whole artificial ecosystem. Fortunately plants usually have a wide tolerance margin and at times suffer torture without complaint, but we should at least do what we can to meet their needs and to give them some guarantee of prolonged safety.

During the first part of this paper I concentrated on aspects of training for the unskilled and semi-skilled gardeners but if we recognize the limitations of knowledge we possess as managers we must be constantly aware that our skilled growers, supervisors and managers need constant updating as we do ourselves. Let no-one say he knows all the answers; and although it is claimed that there is not time to release key staff to visit other botanic gardens, work in the field or attend relevant conferences, this is the only way to motivate the staff on whom we depend so much. Involvement of this nature will increase the individual's confidence, provide a strong professional understanding and help to break down the mistrust that can exist between horticultural and scientific staff.

To conclude, so much depends on basic horticultural skills, and good understanding of procedures, and an outward-looking attitude by key staff including specialist growers. As managers we should regularly review the needs of our staff and give them all the help we can to meet the objectives we are setting for them by our decision to play an active rôle in conservation.

DISCUSSION

J. A. WITT (USA) enquired whether the gardening staff at Kew were trained to deal with the public, since it was inevitable that they would be questioned about their work.

R. I. BEYER replied that no formal training was given, but that staff were expected at all times to be polite, and should pass on any questions they could not answer to a more knowledgeable member of staff.

W. T. STEARN (England) stated that in his many years of involvement with gardens such as Kew, the breakdown in communication was between scientific and gardening staff. Valuable information concerning for example a newly introduced plant did not always find its way to the person growing it.

H. B. RYCROFT (South Africa) said that to aid communications every new member of staff at Kirstenbosch was given a list of all members of staff, and a copy of the institution's last annual report.

B. A. MOLSKI (Warsaw) suggested that there was a lack of communication at an international level and hoped that a journal catering for this need could be evolved. Reciprocal exchange of botanic gardens staff between different countries would also be beneficial.

R. I. BEYER agreed, but remarked that he was most concerned about the understanding required by the man actually handling the endangered plant.

M. AVISHAI (Jerusalem) commented that junior staff required encouragement to supplement their salaries if they were to realize the part they had to play in conserving these plants.

The Rôle of the Media in Conservation

DAVID BELLAMY

University of Durham, England

If one reviews the history of the conservation movement it is impossible to get away from the importance and intimate involvement of television in its development. Sir Peter Scott, Hans and Lotte Hass, Jacques Cousteau, Armand and Michela Dennis and David Attenborough, are household names in many languages. These and many others have been the providers of information to the lay public on a grand scale. As the conservation movement grew so their programmes increased in popularity and changed in emphasis. Conservation in Britain and across the world owes much to these people as it does to the BBC Natural History Unit at Bristol, Anglia Television's Survival Team, Time-Life and a whole host of film makers and publishers.

Perhaps even more important has been the rôle of the news media in spreading the message. Working on the principle that bad news makes good headlines, oil spills, pollution, habitat destruction and overhunting have become as much a part of our news as rape, murder and political unrest. I feel that one of the great hinge points in the whole conservation movement was the Torrey Canyon Disaster and the fact that the whole world was immediately made aware of the slaughter of sea birds and the destruction of other forms of marine life. The other classic was Neil Armstrong's statement from the moon which put the earth into its true perspective—a tiny spaceship fuelled by solar power thanks to the chlorophyll-bearing members of the plant kingdom. This was environmental education on a grand scale and had a great and positive effect on the conservation movement.

It must however be realized right from the start that instant knowledge can cause as many problems as it solves, unless it is handled in the right way. Enormous care must be taken when popularizing anything. Hordes of visitors, however well meaning, may have catastrophic effects on the very thing they are trying to see and are trying to conserve. "To tell or not to tell", that is the paradox of the conservation movement and it is of great relevance in relation to the conservation of rare plants, much more so than in the case of the mobile members of the animal kingdom. Modern 'sponge bag' plant hunters are on the rampage to the most remote corners of the earth and rarity is the commodity they value most.

Having said that I still believe that the public must be informed for the following reasons. An active programme of conservation costs money, public money, and must also curtail freedom of choice to a certain extent; therefore the public have the right to be informed. The answer to this most ingenious paradox is to provide information of the right sort with the right back-up. Here we can learn a lot from the example of the Royal Society for the Protection of Birds (RSPB). Their excellent films, many of which are international award winners, provide the information, stimulating enormous interest, but only when the back-up is already there. This includes well managed reserves and an in-house education programme at all levels. What is more it is self-financing almost to the scale of big business and I use the term in a very positive way.

So to basic knowledge and awareness, and here Botany has a poor record; it is almost as if the subject has lost faith in itself. Botany has been emasculated of much of its practical importance; agricultural botany, horticulture and microbiology are being taught as disciplines in their own right. At the same time the mother subject has itself become changed and diluted by the new frontier disciplines of molecular genetics, theoretical ecology, environmental sciences, etc., so much so that in Britain for example it is very difficult for a student to gain a grounding in classical botany, which is still a cornerstone of many new and important developments. This has had repercussions throughout all levels of education. A realization of the importance of a basic knowledge of plants and classical plant science has slowly but surely slipped away leaving us with all sorts of anomalies. For example, why do our history syllabi spend so much time on the repeal of the Corn Laws while at the same time they find no place for the evolution and development of the crop itself; again why do the same syllabi dwell on the social rather than the biological effects of the Acts of Enclosure, and so on.

The answer at least in part stems from the basic lack of knowledge of plants and their importance, and this is itself due to the fact that botany has disappeared from school curricula at all levels. If educationalists are taxed with the problem, their answer is not that plants are unimportant but that plants are uninteresting, and thus ideal subjects to be swept under the educational carpet. The stark facts are that 'animals do things, plants don't', and so it is therefore easier to make award winning films, instil interest, teach and educate people about animals. For this reason the bulk of natural history television to date has related to animals and birds, and immense public interest has been created in these fields. This has certainly helped the popularity of zoology as a subject and at the same time has pushed botany down the scale of academic interest. The universities and teacher training centres have therefore produced more zoologists, who naturally like to teach zoology and to develop zoologically based curricula, syllabi and exams. So the overall interest in the population of natural historians both amateur and professional has undergone a distinct shift which does not obey the postulates of Hardy Weinberg.

This gap must be closed and television has an important rôle to play, especially now that an increasing number of viewers have the benefit of colour which can only help to make botany come alive. Programmes such as *Botanic Man* made by Thames Television and scheduled for prime viewing time, with its back-up of two books, home learning courses, and extra-mural courses, and *Life on Earth* made by the BBC, are beginning to break new ground, moving in the right direction. The question remains: is Botany and the plant conservation movement ready to capitalize on the initiative?

POSTSCRIPT

It is my belief that the only hope for a meaningful policy of conservation for the many plants now under threat of extinction is to enlist the aid of the world's gardeners. The number of botanic gardens who can carry out 'preservation' work will always be limited as will the number of seed banks. Throughout the climates of the world there exist thousands of keen amateur gardeners who thrive on the challenge of growing rare and difficult plants in their gardens. The initiative is already there in the proposals of the Royal Horticultural Society and those of the newly formed Tradescant Trust to set up a World Gardeners Club. I feel that the weight of all the members of this conference should be placed behind that initiative.

DISCUSSION

S. WAHLBERG (Sweden) asked whether nature programmes on television could be shown without music and suggested for each film several different sound-tracks be made: one of natural sounds, one with added music and one with simply the human voice. He felt that the 'cosmetic of music' misled the public and emphasized that in Sweden nature films were normally shown without music.
D. BELLAMY replied that the musical additions were a decision of the production team, but that most of his programmes were made without music.

A NOTE BY THE EDITORS

Botanic Man is a ten part series David Bellamy has made for Thames Television. The series traces the story of plant evolution and adaptation, showing how plants form the basis of the energy web that sustains life on earth. Extracts from four of the programmes, *Green Print for Life, Living Water, Land of Opportunity,* and *Life on the Limit* were shown to the delegates. *Botanic Man* has been screened in Britain gaining and maintaining record audiences; it has also been sold to many other countries. A lavishly illustrated book accompanying the series has now been published by Hamlyn (Price £6.50); it is also entitled *Botanic Man* and is written by David Bellamy. In addition a project book *Botanic Action* by David Bellamy and Clare Smallman has been published by Hutchinson (£1.95) and a correspondence course called *Green Earth* is available through the National Extension College, Cambridge, England.

PART FOUR

Background Support

Preservation of Plant Resources in Gene Banks within Botanic Gardens

P. A. THOMPSON

Royal Botanic Gardens Kew, Wakehurst Place, Sussex, England

Botanic gardens have a traditional function as centres for the assembly and distribution of plants, sometimes to meet industrial needs, sometimes to display attractive species or forms for amenity horticulture, and sometimes to make available interesting plants for scientific study or education. Recently suggestions have been made that botanic gardens might also play a part in the preservation of plants as a direct aid to the long-term conservation of threatened species or populations. At first sight this appears to be a plausible suggestion, and one which might be expected to produce relatively few problems. However, it must be recognized that this does, in fact, represent an entirely novel situation, and one which in many ways the traditional objectives, standards and techniques of botanic gardens are quite unfitted to fulfil.

Ex situ conservation within botanic gardens is almost invariably less desirable than *in situ* conservation within a natural, or properly managed, ecological continuum, but the suggestion that botanic gardens should take a direct part in conservation must be sympathetically considered. These institutions possess collective resources which must, if properly deployed, be capable of playing an effective part in the conservation of endangered taxa; and often they are uniquely placed to provide advice, guidance and leadership in the development of national policies directed towards plant conservation.

The preference for *in situ* rather than *ex situ* conservation is sometimes posed in an 'either/or' frame, but should not be seen in terms of exclusive alternative strategies at all. There may be situations in which *ex situ* conservation provides the only hope of survival, but apart from this the maintenance of representative populations of endangered or vulnerable species within botanic gardens provides opportunities which may not exist with natural populations, and the two alternatives may be more constructively viewed as mutually complementary activities which can each play an important part in safeguarding particular plant populations.

Plants grown within botanic gardens provide opportunities for:

(a) Study, evaluation and exploitation.

(b) Distribution to other botanic gardens and appropriate research institutes.

(c) Increasing the numbers of individuals by collecting seed from the cultivated populations.

(d) Re-introduction into the wild should the natural populations become eliminated.

These opportunities can be made use of in a cultivated stock (or collection held in a garden) with minimal risk to wild populations, and the cultivated representatives of an endangered taxon may if necessary be sacrificed in order to acquire information on the taxon or population as a whole, in ways which may spare, or contribute to the survival of individuals within, endangered wild populations.

When collections are made from endangered populations it is essential that this should be done in a way which does not significantly hazard the survival of the population. It is also important to ensure that the material collected is capable of fulfilling the purposes for which it is required. Although these two criteria appear almost too obvious to state, traditional methods of plant collecting, which are still widely used by staff of botanic gardens, take very little account of them. Collections of mature plants or established seedlings from small populations inevitably deplete a vulnerable population in a way which does pose a significant loss, especially if repeated, with whatever discretion, by more than one collector. Collections of a few individuals fail to secure a representative population and so minimize the chances of making worthwhile use of the material collected. Collections of numerous plants truly representative of the population are usually impossible or most undesirable for obvious reasons.

There is therefore a conflict between the interests of the vulnerable population in the wild and the interests of the collection to be held and used in a botanic garden, and this cannot be resolved as long as collections are based on established plants or seedlings. It can, very often but not always, be resolved if collections are based on seeds. These are frequently produced in relatively large quantities, the prospects of survival and establishment of individuals are very low, and quite large numbers may often be collected with minimum, or even no significant, hazard to the parent population.

If the collections made are to be used effectively they must, so far as possible, be representative of the original population—and be maintained within the garden in a way which preserves them as a representative sample. These requirements impose severe restraints on the ways in which collections are made, and the ways in which they are maintained, and create demands and standards which are entirely novel within botanic gardens. They present problems which, hitherto, have scarcely been thought about in traditionally maintained collections of living plants, and depend on techniques and approaches which form no part of the training given to most garden staff at any level. Botanic gardens are occasionally referred to as living gene banks, but in practise they fulfil virtually none of the functions of a gene bank, nor, in their traditional form, could they. Since this rôle is a novel one for botanic gardens it is necessary to approach it in a fresh way to try to find solutions independent of established methods.

One way to approach this situation is to start by defining the essential requirements for collection and maintenance of plants to be held as a genetic resource; to use these as a guide to conditions or events which should be avoided, and finally to arrive at a definition of an ideal which may be used both as a guide and a standard.

ESSENTIAL REQUIREMENTS FOR COLLECTION AND MAINTENANCE OF PLANT GENETIC RESOURCES

(a) Representation: Genetic diversity (large population)
(b) Prevention: Genetic erosion (no selection)
(c) Preservation: Genetic integrity (no gene flow)
(d) Retention: Gene frequencies (no distortions to breeding pattern)
(e) Conservation: Long term security (low energy input)

Some of these essentials are deceptively simple. Obviously genetic selection during cultivation is undesirable and should be avoided, but cultivation inevitably involves selective effects. These may be direct, as in cases where a proportion of seeds sown fail to germinate due to dormancy restrictions, or individual plants fail to survive unfavourable weather; or haphazard, in cases where individuals die as the result of accidents—clumsy handling, misdirected use of a hoe, herbicides, etc. Some state requirements about which little information exists. Distortions in breeding patterns affect levels of heterozygosity and occur during cultivation due to the absence of appropriate pollinators and the presence of unnatural ones. Effects of daylength or temperature in an alien environment may be reflected in the sexual development of the plant and the interfertility of different individuals or flowers. Examples of these effects, identified in a relatively few species which have been studied, merely indicate the existence of possible problems in the vast majority of species for which no information is available.

CONDITIONS WHICH SHOULD BE AVOIDED IN COLLECTIONS OF PLANTS HELD AS GENETIC RESOURCES

(a) Small populations (collected or maintained)
(b) Selection (random or directed)
(c) Hybridization (gene flow)
(d) Unnatural breeding patterns (levels of heterozygosity)
(e) High risk survival factors (individual enthusiasms, cultivation in glass-houses, etc.)

Others of these requirements present practical or logistic problems in relation to the normal techniques for cultivating plants in botanic gardens. The need for representative populations indicates a minimum of sixty or more individuals to represent each population. This may not be a problem with small annual herbs—but becomes a major stumbling block with large shrubs or trees within the limited confines of most gardens. The problem is compounded if, as is probable, there is a need to hold collections of several populations of a particular taxon, at sufficient distance one from another to ensure effective isolation and so prevent gene flow between interfertile populations or taxa.

Cultivation under any conditions within a botanic garden includes a high risk to long-term survival—long-term being defined here as a minimum period of fifty years. This risk is clearly increased if populations depend for their survival on elaborate, expensively maintained environments such as glasshouses, or intensively cultivated garden space, or on the dedicated enthusiasm of an individual who is prepared to study and provide for special cultural needs. In each case quite short periods of neglect, changes in staff or alterations in policy practically ensure loss of a proportion or even of entire populations.

IDEAL CONDITIONS FOR COLLECTING AND HOLDING PLANTS AS GENETIC RESOURCES

(a) Single large collection
(b) Full documentation
(c) No regeneration
(d) No maintenance
(e) Constant availability

An ideal situation may never be possible to achieve in practice, but can still serve as a useful standard at which to aim. Rare species are, by definition, unlikely to be available in sufficient quantities to provide for a single large collection. Yet even rare species can have exceptional fruiting years and advantage may be taken of these as and when they occur to collect suitably large and representative samples of seed. 'No maintenance' may be an impossible condition to fulfil in practical terms but some methods of maintenance are much more demanding than others, and some do involve remarkably little annual input of labour, energy or other resources.

Conversely, if it is clearly impossible to create conditions which even begin to meet the standards set, doubts may be raised about the feasibility, credibility or worth of a particular project and this, in turn, could lead to a search for other means of solution. Resources within all botanic gardens are not only finite but usually strictly limited, and little or nothing is to be gained by growing collections, even for conservation, of such dubious quality that they are unlikely to be usable for any worthwhile purpose. Such doubts would be raised by collections made initially from very few plants with a correspondingly narrow genetic base, by collections with little or very inadequate documentation, by collections demanding heated glasshouse accommodation and continuous skilled attention, and so on. If the collections involved were of taxa or populations of exceptional rarity or value it might well be decided to retain them in spite of their inadequacies, or the problems they might pose, but in less exceptional cases the decision might be to go for another collection more complete in itself, or to look for alternative methods or sites for cultivation less demanding of time and energy.

Collections within botanic gardens may be maintained in various ways, each of which poses special problems that need to be considered specifically in relation to the nature and needs of that collection and its maintenance for long-term conservation.

(a) *As plants in living collections*
    Problems of: regeneration—especially annual spp.
                  space—restrictions on number and populations;
                  resources—high maintenance costs in staff time and energy;
                  survival—adverse climatic effects, human error, etc.
                  stability—genetic change from gene flow, erosion, etc.

(b) *As seeds*
    Problems of: survival—maintaining viability;
                  stability—random cytological changes in store.

(c) *As* in vitro *cultures*
    Problems of: technique—maintaining cultures;
                  stability—random genetic change;
                  resources—high staff costs;
                  survival—human error or ignorance.

The use of *in vitro* sterile culture techniques can be ruled out, at least for the present, as a means of long-term preservation of genetic resources of threatened or endangered species. It has the inherent disadvantage of being a technique based almost entirely on clonal propagation from a very few individuals. However, many technical problems still exist which must be solved before the method could be realistically considered, even in exceptional cases, as a possible method of long-term conservation.

Plants growing in living collections pose many problems—summarized above, many of which are soluble given sufficient expenditure of time and money. Their solution might, however, make quite unreasonable demands on the resources of a botanic garden, if a serious effort was made to maintain conservation collections to the high standard required, so as to divert resources from other projects and functions equally important and equally necessary. Difficulties arise from the fact that traditional methods of growing plants and maintaining collections for public display, education or scientific study are insufficiently rigorous and frequently technically inadequate for maintaining gene pools representative of endangered taxa and populations.

This leaves the final alternative of maintaining collections as seeds in seed banks. Experience has now shown that effective methods are available for preparing seeds for storage at sub-zero temperatures of all, or almost all, those species which produce small dry seeds; that deterioration during medium term storage (up to 20 years) is very small, and that seeds may be taken from store and germinated at any time. There are good reasons for believing that the majority of these species, perhaps almost all of them, can be stored for much longer than twenty years—and even for periods measured in centuries if necessary; and that genetic change during storage can be reduced to very low levels indeed. The method gives by far the greatest promise of long-term security combined with low maintenance costs of any of the available alternatives, and, although based on equipment and facilities, is not unduly sensitive to breakdowns or periods of neglect provided simple precautions are taken in the preparation and packeting of the seed stored.

Seed storage suffers from one major disadvantage; it is not applicable to those species which produce large moist seeds, which are intolerant of dessication and so cannot be stored at all at temperatures below freezing point, nor at any other temperature for longer than relatively short periods. Until more research is done on the storage physiology of these recalcitrant species their prospects for long-term storage appear to be nil.

Nevertheless for all other species, maintenance as seeds in cold stores or refrigerated cabinets provides a straightforward, simple and secure method of conservation which can be applied, in conjunction with careful collecting techniques, to fulfil very closely the 'ideal' requirements for a collection listed earlier. Costs may be kept very low by suitable selection of equipment in relation to the number of collections to be held, and input of staff time and other resources may also be kept to a minimum. There is now good reason for positively recommending that, if *ex situ* conservation is to be used for the long-term preservation of a taxon or population, it should be done by making use of seed storage techniques, wherever this is practicable. This may be done in a central facility such as a gene bank which might act in liaison with a number of botanic gardens. Alternatively, and perhaps preferably, each garden might equip itself with the simple, relatively cheap and commercially available equipment required for holding collections of seed in store, which could be used not only in the interests of conservation, but also as a general facility for the effective use of seeds as a part of the total living collections within the garden.

It follows from this that seed storage would be a part of the facilities of the garden—run in conjunction with the living collections, and naturally under the control of the curatorial staff of the garden. Guide-lines are now available, which can be applied to long-term storage needs and for which special scientific training or expertise is not required; in the great majority of gardens therefore there would be no need for the seed storage facilities (or seed bank) to be run by scientific staff. Indeed the problems that do arise are so much a part and parcel of the development of the living collections as a whole—policy, rates of accession, definition of interests, presentation of resources, etc., that there is everything to be said in favour of such banks being fully recognized as a part of the living collections within each garden.

DISCUSSION

W. F. Mearns (Australia) observed that natural evolution stopped with the storage of taxa in a seed bank.

P. A. Thompson stressed that species preservation within seed banks was a last resort.

C. Gómez-Campo (Madrid) stated that seed banks could be the most direct way of preventing the extinction of many species in the tropics where there are high numbers of endemics and where the efforts of conservationists tended to be frustrated by the general lack of legislation and finance and by irresponsible development.

W. T. Stearn (England) indicated the necessity to store seeds relating to cytological studies that had been undertaken.

Setting up a Practical Small Seed Bank*

OLAF OLSEN & FOLMER ARNKLIT

University Botanical Garden, Copenhagen

Before discussing the present system of seed storage at the Botanical Garden of Copenhagen, some brief historical background may be helpful. The Garden itself was founded in 1600, and until the recent establishment of a botanic garden at the University of Århus in Jutland, it was the only university botanic garden in the Danish-speaking part of the country, Kiel in Holstein being the other one in the German-speaking part until 1864. Thus the garden was in a unique position and as a result had many obligations to fulfil. These included the supply of seed, with an emphasis on Danish native plants, to every type of school throughout the country, to be sown in their gardens for educational purposes. As a result, a list of seeds usually available of Danish plants was prepared for distribution to the schools.

Regular seed collection within the garden dates back to the late 1780s. It has been enlarged systematically ever since and has served a dual purpose of plant verification and seed storage.

Every species had its own seed bag which also served as a current reference card, carrying the correct name, authorities, geographical distribution, date of introduction, synonyms and any literature references. Also included was the origin of the plant with its accession number and the annual seed harvest. These bags were called the original seed bags, and all seed of that species, regardless of its origin, was collected and put into the same bag.

Thus from the beginning all records were kept on the seed bags but although records and a card index were introduced later to refer to verified plants within the plant collection, the information on the seed bag was still regularly updated, with the inclusion of notes on the yearly seed harvest, any revision of name, and date of verification. This had the advantage of making all the information readily available during seed checking, so that appropriate action could be taken.

The details of this system can be traced back to Martin Vahl, who became the first lecturer and taxonomist of the garden in 1778, and to F. L. Holbøll, who was curator for 37 years from 1793 to 1829. Holbøll had made a long study tour to Kew before his appointment and here he was in several ways inspired to his future work. Around 1800 Martin Vahl carried out very comprehensive plant verifications and together with Holbøll's clever and skilful insight, the beginning of a broadly based seed collection was formed.

This system was maintained almost until the present day. The harvest of the current year's seed was mixed with that of previous years and the whole mixture

* Read by F. Arnklit

was discarded every five or seven years, generally to coincide with a good seed harvest. Seed from which a freshly verified plant was grown, or seed of known wild origin, was kept in a separate bag or container as original seed to which only collections from the original source could be added.

The seed collection, especially the original seed, has from the beginning been used for the checking of freshly collected seed before the latter is stored or included in the Index Seminum. This annual check is very labour consuming, but is an invaluable safeguard against wrongly harvested or mixed seed. The vast majority of seed, apart from very fine seeds such as of *Begonia* or of orchids, has its own characteristics which can be identified with practice, with the aid of a magnifying glass in cases of doubt. This annual task lasts for about four weeks and is undertaken by two assistant curators.

Up to 1974 two factors caused problems with these accepted principles of seed storage: the necessity to keep seeds in a working room temperature, and the practice of keeping a mixed stock collected over several years in the one bag. Experience had shown that several samples of seed gave differing rates of germination for a variety of reasons, ranging from the age of the seed and the treatment of the seed after harvest, to the conditions of storage. Many annual species which grow in the garden are close to the northern limits of their range and will only set viable seed in years with good ripening conditions. As a result, old seed in some years would have to be used and this might have been of low viability.

More than twenty years ago various methods of seed preparation were adapted from the commercial seed firms. Instead of the normal drying in the open air, the harvest was placed in jute bags which were ventilated in a kiln with regulated warm air. The moisture content was then controlled down to 12–15 per cent. This method facilitated seed cleaning and produced seed of better quality. But the principle of storing the cleaned seed in the uncontrolled temperature of the office or workroom was still very unsatisfactory. This state of affairs, together with the lack of security of this valuable seed collection (which contained approximately 30,000 taxa), made it clear that a change in seed storage arrangements must take place when the administration building of the garden was to be rebuilt.

The background to the then existing seed collection and ideas gained from Kew's seed bank at Wakehurst Place led to the decision that future use of the seed bank should be for two quite distinct purposes. Two similar rooms were built in the new administration building, both safe against water and fire. One room still contains the original type of seed bank, maintained at a temperature of 15°C., and for the time being with a relative humidity of 60 per cent whilst comparative tests on humidity are being carried out with seed samples of different ages. Relative humidity will subsequently be reduced to 30 per cent. This seed bank will be used in future for the storage of original seed samples and for checking the annual seed harvest, as well as for scientific purposes such as seed identification. The storage of seed in the original seed bags will be continued until the second room, as described below, has been established.

The second room will contain the seed bank proper. Bearing in mind ease of use, variation of temperature, and the garden's estimated capacity, we intend to install large deep freezers with differing temperatures rather than separate cold store rooms. The first deep freezer has already been installed and will cover temperatures down to −20°C.; two more will follow, one at −10°C. and one at +5°C. This temperature range is intended to cover the demands of the many different species from the various climatic zones which grow outside or in the glasshouses of the garden. The freezer already installed has adjustable shelves, on which stand a number of aluminium trays. Each tray can hold 33 containers of 50 ml or 24 containers of 100 ml, giving a total capacity of 2640 and 1920 containers respectively. For the larger amounts of seed 250 ml and 500 ml containers will be used. These containers will be wide-necked glass jars with screw caps. This first deep freezer will be used temporarily for the long-term storage of seed for gardens use. As far as possible we will collect seed of known origin and preferably of natural source in large enough quantities to ensure sufficient seed for a number of years' sowing.

Seed collected from a particularly good harvest will also be stored. The general policy is to concentrate firstly on the collection of seed of annual species (there are about 2400 such species of both native and foreign origin grown either in the open or under glass), provided that these will tolerate low temperature storage, and secondly on seed of Danish plants, either annual or perennial, collected from the wild. When a successful pattern has been established we will define and fulfil our further needs.

The first deep freezer will suffice for these initial steps. The other two deep freezers are intended for the storage of seed of tropical species and the seed of certain trees that will not tolerate storage at −20°C. The intention is for annual regeneration to maintain the collection of both biennials and perennials both under glass and outdoors.

The traditional emphasis on the cultivation of native Danish plants has been continued since the founding of the present garden in 1874, and a proportion of the garden is devoted to these plants, all of which were collected from within Denmark. Details of their origin are entered in the garden's accession books. The emphasis of this conference on protecting and preserving threatened species is also reflected in some of the activities of the botanic garden and of certain natural history societies, who are working in conjunction with each other to collect seeds of as wide a representation as possible of the Danish flora. These are included in our Index Seminum, although the ultimate intention is of long-term storage in our seed bank. The same procedure will be applied to the harvest and storage of seeds from Greenland, with a similar emphasis on rare or isolated endemics.

A seed drying room has been built attached to the seed bank, whilst a germination testing laboratory is currently being built to monitor seed viability of both old and recently collected seed. This laboratory can be used by the botanical institutes at the University for seed work, and we also intend to work on seed that is difficult to germinate. All seed harvested will be dried in a warm air kiln down to a moisture content of between 12 and 15 per cent. Seed intended for banking will be dried further to about 2–3 per cent to secure the best possible storage.

The seed bank at Copenhagen Botanical Garden will not exist in isolation, because we are linked through an agreement with the Danish Agricultural University to the newly established Nordic Gene Bank in Lund, Sweden, for mutual co-operation. Emphasis will be placed on species from the Danish flora which might be of special interest in future breeding work.

A problem of special interest is the number of plants of a certain species which can be held in a botanic garden in relation to the space available. Another is the maintenance of seed from very rare plants after their storage in the seed bank over a number of years. It was calculated in the first case that at least 50 individual plants were needed to produce seed for banking in order to avoid in-breeding and limitations in a naturally heterogeneous cross-pollinating subject. In the second case it is reasonable to bulk up seed by harvesting from a suitable sowing in open ground, to be carried out in our trial grounds outside Copenhagen.

This paper began by discussing the twin uses of the seed bags for seed storage and for index cards; these will ultimately be linked to a computer as are the systems of Kew and Edinburgh. Our intention is that every taxon stored will have its own individual accession number, linked to other plant information within the garden to provide instant information retrieval instead of the slow procedures of the past.

DISCUSSION

S. WAHLBERG (Sweden) emphasized that the preservation of a plant in a seed bank or a botanic garden was a last resort of conservation and that this was no substitute for conservation in the wild.

F. ARNKLIT sympathized with these views, but nevertheless he felt that a seed bank made a useful contribution to the aims of conservation.

The Potential and Progress of the Technical Propagation Unit at the Royal Botanic Gardens, Kew

HARRY TOWNSEND

Royal Botanic Gardens, Kew, England

The justification for building the Technical Propagation Unit for work on *in vitro* culture can be better appreciated by reviewing the plant introduction procedures at Kew, which had varied little from early days until very recently. The number of taxa maintained in cultivation had risen only marginally since the turn of the last century, despite the constant stream of often haphazard plant accessions; space considerations aside, this was almost certainly due to a lack of appreciation of the techniques required for both establishment and regeneration. Generally, the condition of plant material on arrival was poor, due to insufficient knowledge, unsatisfactory packing and unreliable transport; and since field notes were generally not available the propagator was not able to appreciate the needs of the plant. Thus plants and seeds were usually given a 'standard' treatment which was governed by their broadly alpine, temperate or tropical origins, so that the plants which survived were usually tolerant of considerable abuse. The majority of botanic gardens worked on similar lines, so that these plants formed the basis of a 'botanic gardens flora', whilst very many rare and unique plants failed to grow. The only plants in this latter category which survived were either those whose requirements were exactly matched by the environment in which they suddenly found themselves (a remote coincidence), or those plants which were the special interest of a particular grower who over the years had built up a deep and personal knowledge of the specific requirements of a group of often very 'difficult' plants. Such collections would invariably collapse when that person left.

Nowadays collectors in the field collect on the basis of specific requests from botanists or horticulturists. Improved storage techniques and more efficient transport allied to meticulous field notes mean that the majority of plants on arrival at Kew have the potential for growth. Yet until well within the past decade, despite the improving background knowledge, progagation facilities at Kew were still no more than a sophistication of methods used for a century or more and it was recognized that the resources were lacking to cater for material which was often unique and consequently of conservation importance. Thus for the first time the concept of providing a facility for *in vitro* techniques as a service within the Living Collections Division was accepted, and this led to the development of the Technical Propagation Unit in 1973.

Many universities and research institutes had used tissue culture techniques and had studied seed germination under laboratory conditions. They had applied these for often limited research purposes, whereas commercial firms were applying these techniques to specific horticultural problems. This was the first time, however, that such techniques were being adapted to attempt the propagation of an unlimited range of rare plants, whose propagation by conventional methods was not possible but whose importance in the field of plant conservation was great. Furthermore it was not only essential to propagate from material of natural source origin, but it was necessary to guarantee genetic purity; this meant a greater emphasis was placed on propagation from seed than from vegetative material.

No comparable knowledge existed on which plans for this new unit could be based. Nevertheless we were grateful to have been able to visit and receive advice from scientists at the University of Leicester, the University of Nottingham, the John Innes Institute, the National Vegetable Research Station, the Glasshouse Crops Research Institute, Twyford Laboratories Ltd and Unilever Research Laboratories whilst our plans were being formulated, and for their constant help and encouragement since then.

Any mystique is entirely of other people's making for the normal horticultural practices are used, although with more precision and under sterile conditions; it is important to think of this unit as more relevant to propagation than to a laboratory, and this helps to put its purpose into a correct perspective. Thus the test tube compares with the closed frame; the incubator, although more precise, compares with the heated glasshouse; whilst the precisely defined medium is a sophistication of the cutting compost. Staff were originally selected from horticultural workers at Kew who had background laboratory experience and some scientific knowledge; this meant that they were able to apply their practical horticultural knowledge to many problems in a way which might not be immediately evident to the pure research scientist. It is important to realize from this that the Unit is not intended to undertake original research, although developmental research is inevitable when dealing with plants of whose propagation requirements nothing is known. Every step is recorded for reference and where justified for publication. This information is filed in a data bank which contains information about the propagation of any and every plant, and which is being constantly built up from regular literature searches and through close contact with the Physiology Section under P. A. Thompson at Wakehurst Place.

The most definitive emphasis of the practical intent of the Technical Propagation Unit is provided by its status as a service unit for all the living collections grown at Kew; each request for work must be assessed in terms of its relevance to the work of the Royal Botanic Gardens as well as for the probable work commitment involved, since space and staff are limited and priorities must be clearly established. No plant should be considered unless it is rare, of conservation importance or recently introduced; it should be of natural source origin, preferably in the form of seed; and it must be difficult to propagate by conventional methods. Exceptions to these criteria are few: they include such purposes as the bulking up of plants for research purposes, or the improvement in health of heavily virus-infected plants for cytological study. Surplus propagules are offered to other botanic gardens or scientific institutions as a part of the normal policy of Kew, and distribution *in vitro* has enabled the stringent quarantine regulations of many countries to be satisfied.

Many problems were encountered initially as the unit was planned and implemented, and scientific techniques have been adapted to meet these in many cases. It was thought at first, for instance, that it was necessary to supply filter-sterilized air to the transfer rooms and to maintain a positive pressure in the rooms in which sterile material was handled so that all airborne micro-organisms should be totally excluded. It was also thought that people should be dressed, gloved and masked in sterile clothes whenever they came into contact with any plant material in the unit. However, experience indicates that these precautions were not essential and it was realized that with normal care and correct techniques the only place in which filtered air was essential was the laminar airflow benches in the transfer rooms, and that no special clothing (apart from recommended hair covering) or gloves were needed to obtain a very low percentage (between 1 per cent and 3 per cent) of contamination in the test tubes, which is perfectly acceptable; the expense and personal discomfort involved in lowering this rate could not be justified.

TECHNIQUES

In each case the data bank is consulted before any work begins, but since little propagation or tissue culture information is recorded elsewhere about the majority of the plants which come within the ambit of the Technical Propagation Unit, methods used are largely empirical. All plant material, whether seed, spore or vegetative, is sterilized in a solution of bleach (sodium hypochlorite) with an added wetting agent; the duration of immersion before washing in distilled water depends entirely on the sensitivity of the plant material involved. Success in every case can only be achieved by trial and error, and the success rate can be increased by the use of clean starting material.

Growth from seed is always to be preferred, since the possibility of genetic mutation is minimized. This is a greater possibility when tissue culture is used, particularly if there is an intermediate callus stage. However, should viable natural source seed not be obtainable it is necessary to attempt tissue culture, often from axillary and terminal buds but on occasions using any other part of the plant up to the ovary, stamen or even petal of the flower.

The 'recipes' for a great many media have been published and variations have been postulated for a diversity of purposes, but the basic medium used at Kew is that formulated by Murashige and Skoog, to which hormones, carbohydrates and agar are added as required.

After the medium has been made up, it is poured into the test tubes which are then capped and autoclaved (pressure steam sterilized) to ensure sterility. These are then used as soon as possible. The dissected vegetative material or dissected seeds are 'planted' in the tubes within the sterile environment of a laminar airflow cabinet, the seeds first being surface sterilized as described above. The racks of capped and 'planted' test tubes are then placed on the growth shelves for which warmth and light are provided. Controlled environment cabinets are extremely expensive and it was necessary to produce our own growth shelves for this purpose, in which the temperature of the propagule rather than that of the environment is controlled by the cooling of the shelf on

which the racks stand. Cooling is necessary since the room in which the growth
shelves are situated is overheated as a result of the power used by the electrical
equipment (particularly the lighting); this is done by cooled water circulated in
pipes in contact with the shelf on which the racks of test tubes stand. The
temperature at propagule level is monitored by a resistance thermometer and
recorder, and a thermostat controls the water temperature in the shelves. This
system has been adapted from a method seen at the John Innes Institute.
However these are doctrines of perfection, for differentiated growth from seed
or from vegetative material cannot be guaranteed. While we think it true to say
that every cell is totipotent (having the potential for differentiated growth) and
that therefore every plant can be propagated vegetatively as well as from seed,
it is necessary for all but a very few particularly amenable species to find out
how this can be done. There is no doubt that propagation from seed brings out
the best chance of success.

Seed is sown on an agar medium without additives in test tubes, and if it does
not germinate under standard treatment on the growth shelves it is placed in one
or more of a range of incubators kept at controlled temperatures. These are
initially constant, with a steady 12 hour day and night cycle; but if this quite
wide range fails to induce germination, and if viability tests (on those seeds large
enough to do so) indicate that the potential for growth exists, then a variety of
precise temperature variations are used, ranging from an initial period of cold
storage followed by standard temperatures, to fluctuating day and night temperatures with the variation of total darkness. Germination inhibitors exist in
many seeds, having evolved for specific climatic conditions; thus in some cases
it is necessary to excise the embryo, to remove the testa or to remove the immature
seed from the green pod and to attempt to culture these *in vitro*.

The sowing of seed of epiphytic orchids, without being able to undertake
work in the very important field of mycorrhizal association, and without any
variations of light or temperature, using a constant temperature range of between
20°C. and 25°C., has produced a success rate of approximately 30 per cent.
The germination of all other seed has been considerably higher; in particular
the germination of fern spores, because of the considerable amount of work
undertaken on relatively few tropical species to date, has risen from an original
low success rate, principally through the use of spores before dehiscence, to
approximately 70 per cent.

Vegetative propagation, simply because each species or even each clone might
demand a different temperature or a different hormone balance to the medium,
gives a low rate of success; the range of variation within these parameters is
almost infinite. Unless one starts with information drawn from the data bank
(which is itself often incomplete, often omitting for instance such information
as the time of year, an essential factor since greater success is achieved if one
works with the natural growth rhythm of the plant), work must be completely
empirical. Thus *any* success with the vegetative propagation of the rare or
endangered plants with which we work is encouraging; and a success rate of 5
per cent for woody plants rising to 20 per cent for other plants could be regarded
as a high estimate.

Problems have not finished when differentiated growth has been obtained, for
the transfer of the young plant to compost, and the subsequent weaning, pose

further considerable problems. Roots produced *in vitro* are invariably thick and fleshy, rather than the fibrous root system produced by the majority of other plants in the ground; and the breaking of the root cap of the only root means death to the young plant which generally has no power to produce more roots. This is an aspect which needs greater study and such intermediate stages as cultivation in modified medium, or the introduction of additional light, or the taking of cuttings from shoots produced *in vitro* are factors which should be considered. Thus great care is always taken when removing the propagule from the test tube and when planting it in open, well drained compost either under an intermittent mist system or in a humid propagating frame, from where it must be weaned with care.

SUCCESSES

Rare plants raised from seed include not only many species of epiphytic orchids and one species of British terrestrial orchid, *Dactylorhiza fuchsii*, but such plants as *Aloe polyphylla*, *Chiranthodendron pentadactylon*, several species of *Vellozia* from South America, *Muiria hortensae* and *Musschia wollastonii*. Spores of twelve species of tropical fern have so far been sown, none of which had previously germinated in cultivation at Kew; of these eight have formed prothalli and four (*Dicranopteris linearis*, *Platycerium coronarium*, *Pneumatopteris callosa* and a *Sphenomeris* sp. collected recently on Mount Roraima) have produced plants ready for addition to the Kew collection and for distribution. Amongst the wide range of other difficult plants raised from seed are several species of *Drosera* and *Nepenthes*.

Many rare and often unique plants of conservation interest from within the Kew collections have been propagated vegetatively, often from vestigial material; particularly pleasing has been the successful propagation and distribution of the rare *Solanum lidii* from the Canary Islands. Work on the propagation of cycads has produced differential growth of *Zamia pumila* and callus formation on 11 of the 16 species attempted, whilst work on the vegetative propagation of clones of Cocoa (*Theobroma cacao*) on behalf of the Quarantine Unit at Kew has given cause for cautious optimism as a result of early success. As a result of this success there is no reason why the Technical Propagation Unit should change its function of a service unit for the Royal Botanic Gardens, Kew. It is also hoped that such units will become increasingly widespread in botanic gardens for the essential propagation of endangered plants.

ACKNOWLEDGEMENTS

I would like to thank A. M. Chabert and P. L. Gibbon for their help with the writing of this paper.

The Organization for the Phyto-Taxonomic Investigation of the Mediterranean Area (OPTIMA)

César Gómez-Campo

Universidad Politécnica, Madrid

The origin of OPTIMA (Organization for the Phyto-Taxonomic Investigation of the Mediterranean Area) dates back to 1974. With the *Flora Europaea* project already very mature, a new project on *Flora Mediterranea* was proposed as a possibility for continuing a similar line of international botanical co-operation. Very soon, however, a realistic point of view emerged in the sense that such a project was not feasible for the moment. The gaps in knowledge were so apparent and so abundant for many local areas and also for many taxonomic groups in the Mediterranean region that the writing of a Flora should wait until a more mature stage could be reached. In fact the real immediate need was to stimulate floristic research in the area. OPTIMA was born primarily to fulfil this aim.

The richness and diversity of the Mediterranean flora is well known and very easy to demonstrate with numbers. For instance, four out of every five European endemics grow in the Mediterranean region. While the number of endemics in the central or northern European countries is usually counted in tens or less, in the Mediterranean countries they need to be counted by hundreds.

The causes of this situation are very complex. In short, two recent geological events—quaternary ice ages and the desertification of the Sahara—trapped the Mediterranean area between them; its flora could not only escape from the devastating effects of both events, but often experienced recent evolutionary diversification as a consequence of the abundance of suitable niches and micro-habitats.

For OPTIMA, the Mediterranean region is taken in a wide sense to include some of the adjoining areas. One of these is the Irano-Turanian region where several Mediterranean groups have their centres of origin; another is the Macaronesian islands where many relicts from the Tertiary Mediterranean flora still survive. The same broad view is applied to taxonomic groups. At least theoretically, non-flowering plants are also included in the field of interest, although for the time being less attention is being paid to them in comparison to flowering plants.

Emphasis is put on the involvement of local botanists in their own particular problems. To become an OPTIMA member, it is not necessary to live in the area. To work or simply to be interested in the Mediterranean flora is sufficient.

The initial subject, taxonomy, soon became enlarged to cover other related disciplines, very especially conservation which is so important in the Mediterranean region, not only because there are so many things to conserve, but also because it involves several major problems that to a large extent are specific to

195

the area. Mediterranean lands have been historically affected by human activities since ancient times, much before other areas as for instance in central Europe. It can be easily imagined how many events had already occurred around the Roman 'Mare Nostrum' before the first devastating agent for vegetation—the hoof of Attila's horse?!—began acting in central Europe. Another serious problem is the occurrence of hot and dry weather during several months of summer. This applies an important physical limiting factor on the recovery of any damage done to the vegetation.

In many Mediterranean countries, industrialization is either well advanced or being developed through the same steps and patterns that have already been demonstrated to have devastating effects on nature in other countries. Finally, tourism is an important present or potential activity for most Mediterranean nations. Tourism has a curious doubled-edged effect on nature. On one side it is positive as it stimulates the appreciation of natural values. On the other, it frequently happens that irrational tourist developments compete with nature for the best natural areas. All these are good reasons why OPTIMA is putting much emphasis on conservation.

By early 1978, OPTIMA had more than 500 members. An executive council of nine members plus an international board of sixteen members constitute the ruling bodies. The first President was G. Moggi from the University of Florence; the present President is Professor S. Rivas Martínez from the University of Madrid. The Secretary, an important person for the organization, is Professor W. Greuter, who was formerly in the Conservatoire at Geneva and is now in the Berlin-Dahlem Botanical Museum. Most OPTIMA activities are managed by a number of commissions which are formed by a variable number of members, usually from six to twelve. These commissions are:

1. The 'Current Research' commission, which is preparing an inventory of research projects being developed, in order to secure proper co-ordination.

2. The 'Floras and Monographs' commission intends to stimulate the writing of Floras and to announce and co-ordinate botanical expeditions.

3. The 'Plant Resources and Conservation' commission is now preparing lists of rare, vulnerable and endangered Mediterranean plant species in close connection with the TPC (Threatened Plants Committee of IUCN).

4. The 'Seed and Living Material' commission is trying to stimulate the development of a network of seed banks of endemics, aimed at conservation and also at improving the availability of rare plant material in living form for research purposes.

5. The 'Publications' commission produces a Newsletter twice a year. This is proving to be enormously useful for mutual communication between OPTIMA members. Also, bound collections of reprints concerning the Mediterranean flora are circulated under the name of *OPTIMA Leaflets*. The printing and /or distribution of other relevant publications is also promoted, partly through the negotiation of discounts for members with publishers and dealers.

6. The 'Prize' commission awards a prize every three years to works of merit related to the Mediterranean flora.

7. A Commission for mapping the orchids in the Mediterranean region has recently been created.

OPTIMA has celebrated two successive meetings, the first in 1975 in Iráklion (Crete) and the second in 1977 in Florence (Italy). A third meeting is being planned for 1980 in Ljubljana (Yugoslavia).

Physical and Chemical Soil Factors Affecting the Growth and Cultivation of Endemic Plants

J. A. VARLEY

Tropical Soils Unit, Land Resources Development Centre, Reading, England

SUMMARY

Many plants will grow over a wide range of soils with greatly differing physical and chemical conditions, often becoming modified or specialized when extremes are encountered. Usually, little is known about the conditions best suited for particular plant species that are of little or no economic importance to man.

A number of soil conditions are discussed, and suggestions made regarding the principal factors required for the well-being of all plants, including those that are rare or endangered.

The specific example of *Trochetia erythroxylon*, the St Helena Redwood, is used to illustrate the practical aspects of conservation and the deductions that had to be made to save this endangered species.

INTRODUCTION

Many plants grow well, flower and produce seed in a very wide range of physical and chemical soil conditions. During the past few years my laboratory has analysed nearly 25,000 soil samples collected throughout the tropics. In certain parts of the world some of these soils contain abnormally high levels of elements such as chromium and nickel, not generally regarded as necessary for the normal growth of most plants. From aerial photographs and ground observations many different plant species are seen to grow on these unusual soils, and the questions can be asked, "Do these plants grow there because of or in spite of these elements?" or "If these plants were removed from their natural environment, would they have to be supplied with elements such as chromium and nickel in order to keep healthy?" At present, I do not believe that anyone can give a definite answer to these questions. Today there is probably no urgency in giving an answer anyway, but if in the future any of the plants growing on these abnormal soils were to become endangered species, then an answer might become very important.

What then are the major factors that could affect the growth of plants in cultivation? Six are of immediate interest:

1. Soil pH—the acidity or alkalinity of a soil.
2. Electrical conductivity (EC)—the quantity of soluble salts in the soil moisture.

199

3. Soil texture—air and moisture spaces between the soil particles.
4. Water quality—the amount and type of soluble salts.
5. Type of fertilizer.
6. Bench sand or bench covering.

SOIL pH

Probably the most important single factor affecting plant growth is the pH of the soil. Cabbages and potatoes are found in most home vegetable gardens throughout Europe, and if there are no pathological or entomological problems, a reasonable crop is usually obtained over a very wide range of soil conditions. This is not true of rhododendrons and azaleas, which prefer acid conditions; when the soil tends towards alkalinity then lime-induced interveinal chlorosis results.

Soil pH can also be an important factor in the symbiotic associations of plants. Such species are often quite specific in the conditions they require for best growth. One half of the partnership may be able to grow over a wide range of soil pH, but the other half may be very specific in its requirements. An example of this phenomenon is found in the Leguminosae. Generally plants of this family grow best in acid but near to neutral conditions, around pH 6.0-6.5. The soil pH for best growth is not dictated by the legume but by its symbiotic partner, the rhizobia associated with the root nodules and responsible for the fixation of atmospheric nitrogen. In order to be effective in fixing nitrogen the rhizobia require a very small but constant supply of molybdenum. In acid soil conditions the availability of molybdenum is very low, but this increases as soil pH rises. Near to neutral conditions the rhizobia are able to obtain sufficient molybdenum to become efficient fixers of nitrogen. In turn the legume grows well since it is able to obtain its nitrogen requirements in a readily usable form from the rhizobia.

ELECTRICAL CONDUCTIVITY (EC)

This is a measure of the amount of soluble salts in the interparticulate spaces within the soil. Plants obtain their essential elements from solution in the soil moisture. Such solutions are, however, extremely weak.

In general, plants do not tolerate high salt levels in the soil solution, and in particular object to high sodium levels. There are exceptions such as some species which grow in extremely saline conditions.

Giving plants too much care and attention can cause just as many problems as neglecting them. High salt levels may arise due to the over-feeding of soluble fertilizers. Plants growing in pots that are not regularly leached to remove excess soluble salts often show poor growth or even die because of the presence of a high concentration of these salts in the soil solution. In the small soil volume of a pot, the salts become more concentrated as the soil dries out and this is an important factor when considering the frequency of watering. Salts accumulate in different parts of a pot dependent upon whether it has been watered from the top or the bottom. Table 1 illustrates this point.

TABLE 1

Electrical conductivity of soil in top and base watered pots (m.mhos 1: $5H_2O$)

	Top Third	*Middle Third*	*Bottom Third*
Top watered pot	0.28	0.43	0.84
Base watered pot	0.52	0.21	0.21

SOIL TEXTURE

In general plants grow better in well-aerated soils. Heavy compact soils cause restriction of root growth which subsequently leads to poor top growth. Heavy soils are also more inclined to suffer from poor drainage or waterlogging which in turn causes the elements iron, manganese, nickel and chromium to be mobilized into a soluble form, and this can cause the death of some plants.

Extreme requirements of root aeration can be found. At one end of the spectrum, many species of Orchidaceae must have aerial roots if they are to make any growth at all, while at the other end rice grows and yields well in waterlogged conditions, having alternative arrangements for the aeration of its roots.

I have recently illustrated by pot trial the extreme aeration requirements of one species of orchid, *Epidendrum radicans*. For the past three years I have grown this orchid in a cool greenhouse. Given light, a little warmth and moisture, it will grow very well in most suburban houses, giving flowers only 4–6 months after transplanting, and thereafter flowering throughout most of the year. The secret of success is aeration of the root system. In my laboratory a number of similar cuttings were taken, each 15–20 cm long with three roots of about 10 cm. The plants were divided into two groups, one with all the roots buried in a very open textured peat/sand mixture, and the other with two roots buried in the same mixture but the third root left in the air. Treatments and conditions were the same for the next six months. Even though the soil was of the lightest possible texture, all of the first group became moribund or died while all of the second group flowered after 4–6 months.

WATER QUALITY

Plants growing in a suitable acid soil when first planted may eventually grow badly or die if they are watered with hard water for any length of time—that is water containing calcium or magnesium bicarbonates. These elements replace the hydrogen ions in the original acid soil, making the soil progressively more alkaline. Genera that are known calcifuges and prefer acid soil conditions should only be watered with rain water or de-mineralized water. Naturally, this is going to be more expensive than using tap water, but this is surely cheaper than having to mount new expeditions to collect more plants in order to replace the ones that have died.

TYPE OF FERTILIZER

Including minor elements, plants are known to require at least twelve elements in order to achieve good growth.

Basically, the choice of fertilizer depends on two factors:

1. Is the plant a calcifuge?
2. What residual reaction is required?

In the case of nitrogen, for example, three common fertilizers may be used—ammonium sulphate, urea and calcium ammonium nitrate. Ammonium sulphate produces an acid final reaction, reducing pH, and is clearly suitable for calcifuges. Urea is neutral in reaction, but with a tendency to produce an initial alkaline reaction due to the release of some ammonia gas. Calcium ammonium nitrate will obviously have a residual alkaline reaction due to the calcium and is of no use for calcifuges. Fertilizers containing ammonium salts should never be applied to alkaline soils as the ammonia gas released is harmful to plant roots. The same principle applies to all other fertilizers. For example, potassium may be supplied as the very acid potassium dihydrogen phosphate, or the less acid di-potassium hydrogen phosphate, or the neutral potassium chloride or sulphate.

Often when all else fails, ailing plants are given a 'trace element tonic'. Thought should be given before resorting to this as some plants differentially remove one element by a scavenging action. A well known case of this is shown by two plants that often grow side by side in Africa and the Far East: the rubber tree, *Hevea brasiliensis*, scavenges boron and is rapidly killed by only small amounts of this element, but is able to withstand massive doses of copper. On the other hand the oil palm, *Elaeis guineënsis*, is unaffected by quite large dressings of sodium borate, even in the leaf axils, but the leaves rapidly become necrotic when sprayed with only a 1 ppm copper solution.

BENCH SAND

In many greenhouses the benches are covered with a 2–4 cm thick layer of coarse sand in order to maintain a moist atmosphere around the pots and plants. Rarely is the source or specification of the sand considered, and it may contain as much as 25 per cent calcium carbonate. Under constant moisture conditions and with carbon dioxide in the atmosphere, some of this calcium carbonate may be changed into the soluble calcium bicarbonate, and this alkaline product may then find its way into a pot by capillary action and so eventually raise the soil pH. As a matter of principle all bench sand should be acid. No soluble alkaline by-products are then formed that could affect the growth of calcifuges.

Some five years ago my daughter bought a Venus Fly-trap, *Dionaea muscipula*. Its natural habitat is the extremely acid swamps of the southeast United States, where the available nitrogen is nil and the plant nitrogen source is from the protein of insects. For best growth the plant must be grown in waterlogged acid peat. It would surely have died had I not instructed her to use only distilled or rain water rather than the local tap water which contained very large amounts of calcium and which would eventually have raised the pH of the soil.

Botanic gardens are also interested in rare, exotic or endangered species and can finance expeditions to collect such plants. There are certainly no growing instructions stuck to a rock face containing a rare plant, just as there were no instructions on the Venus Fly-trap. I feel that all too often a botanic garden, while still in a state of euphoria at having obtained a rare plant, will place it in that panacea for all growing ills—the commercial growing compost. I do not wish to belittle this type of planting medium in any way. It is a good sterile average textured soil, suitable for average plants, grown by the average gardener—*but it is not suitable for all plants.*

THE CASE OF TROCHETIA ERYTHROXYLON

The problem of providing the correct soil conditions to obtain the best growth was most forcefully brought home to me during a visit to the island of St Helena. The Agricultural Officer explained that the shrub *Trochetia erythroxylon*, known locally as the St Helena Redwood, was in great danger of becoming extinct in the very near future. In May 1976 there were only two specimens remaining; one was growing on an exposed position high up (790 m), in one of the small remaining areas of original forest. Due to heavy battering by the wind, this plant was in danger of dying and in any case did not appear to be in any condition to flower. The other plant, a 'tame' specimen, was growing in a relatively sheltered position outside the forestry office. This plant flowered each January and set seed which ripened in April. Each year the seeds were collected and easily germinated. Two or three months after germination, when the plants were about 10 cm tall, they were given to various amateur gardeners for growing in sheltered positions in their own gardens. Invariably the plants became moribund and died.

In 1976 I was given some fresh seeds from the one plant still producing them. I also collected soil from one of the original forest areas. Analysis of this soil showed it to be very acid and to contain a high percentage of organic matter, suggesting that the plant could be a calcifuge. Based on the analytical data a synthetic soil was made up and in this germination and preliminary growth of *Trochetia erythroxylon* were excellent. Only distilled water and chemicals having an acid reaction were used for feeding. After two years the plants were 120–180 cm tall and were given to the Royal Botanic Gardens, Kew, where they are at present flourishing.

The question that must be asked now is, "Why should a plant, indigenous only to St Helena, where it has evolved over several million years, no longer grow on the island of its origin?" The answer lies in the recent history of the island and the proof is adequately supplied by soil and water analyses.

St Helena was discovered by the Portuguese in 1502. It was said "to be richly clothed in vegetation with gumwoods and other indigenous trees overhanging some of the sea precipices and to possess an abundance of water". The Portuguese introduced European vegetables, trees including citrus, and goats. They also used the tall forest trees for ship masts and planks. In their turn the Dutch and eventually the English colonists continued the policy of felling forests, and allowing goats to breed naturally to ensure abundant supplies of fresh meat. After only about 40 to 50 years, removal of the forest and the prevention of its regeneration due to the goats eating the seedlings, had caused much erosion with massive removal of top soil from the periferal areas of the island. Many orders for goat control were made—but too late. The acid top soil, developed over millions of years, was lost in ever-increasing quantities as the forest was removed. Today, of the total land area of 11,900 hectares, 8,300 hectares, mostly below the 1500 ft (457 m) contour, are barren.

St Helena is a volcanic island; the last eruption took place some 8 million years ago. The bulk of the basaltic shield is formed of lava and inter-bedded pyroclastics with later eruptions adding trachyandesitic lavas and some highly alkaline dyke intrusions. Thus the rocks underlying the acid soil are alkaline, often containing up to 7 per cent sodium. Any new soil that had been formed during the last 350 years in the eroded areas has tended to be alkaline in reaction. With the ever-decreasing annual rainfall brought about by the uplift of air over the barren areas, the soil was not well leached and has tended to remain alkaline.

A comparison of two soils is given in Table 2. One soil is from the remaining indigenous forest and the other from an eroded area. The important parameters are the high pH and exchangeable sodium of the soil from the eroded area.

TABLE 2

Comparison of the analyses of soils from indigenous forest area and eroded area on St Helena

	Indigenous forest	Eroded area
pH (1 : 5 H$_2$O)	4.1	7.8
EC (1 : 5 H$_2$O) (m.mhos)	0.12	0.24
Organic matter (per cent)	50	3
Exch. sodium (m.eq/100g)	nil	2.2
Rainfall (mm)	1200	530

Table 3 gives the overall picture of the increasing pH with decreasing altitude. Samples taken from the lowest altitude show lowest rainfall and greatest cation saturation which in turn is related to a high pH.

TABLE 3

pH, rainfall and percentage saturation variation with altitude on St Helena

Ht(m)	pH (1 : 5H₂O)	Average rainfall (mm)	Percent Cation saturation
800	4.1	1200	ND
610	4.4	970	4
560	5.2	870	60
500	6.1	800	81
450	6.3	610	83
260	7.8	530	100
180	8.1	300	100

Trochetia erythroxylon was once abundant on St Helena. It was classed as a middle altitude (250–550 m) plant. Attempts to establish it again in this range were totally unsuccessful, since the soil pH is today much higher than 350 years ago when man started to change the environment.

The mystery of why this plant will no longer grow on the soils of the middle altitude of the island has now been solved: the seedlings had been given to gardeners whose land was partially eroded and where the soil pH was no longer suitable for the growth of this calcifuge. In this case I was extremely lucky in that the history of St Helena, since man first arrived in 1502, is well documented and many of these writings are still available. These first-hand accounts of previous vegetation, lack of goat control and the advance of erosion, together with modern soil analysis, has helped to solve a problem that has baffled many people for a number of years. I hope it has also saved the shrub *Trochetia erythroxylon* from extinction.

RECOMMENDATION

In order to ensure the best conditions for the growth of a rare or endangered species, soil samples should be collected together with the plants, one from the surface (0–15 cm) and one from the subsoil (15–45 cm). Subsequent analysis should provide the information necessary to prepare a suitable synthetic soil.

After obtaining the soil analyses one must now ask the question, "Is the plant rare or endangered because its growth requirements are ultra-specific or is it because the overall environment has changed?" If the latter applies then the soil analysis only reflects the present, and attempts must then be made to extrapolate backwards in time in order to see what the conditions may have been like when the plant was more plentiful.

PART FIVE

Special Groups

Botanic Gardens and the Conservation of European Orchids

INGRID VON RAMIN

Palmengarten, Frankfurt, Federal Republic of Germany

The world population is growing continuously. Industries are expanding and swallowing vast areas of countryside. Is it necessary that animals and plants are exterminated and destroyed? People draw profit and pleasure from nature and by threatening wild creatures it makes us all the poorer. What is the use of a fine television set if it cannot reproduce the rare beauties of nature because they were superseded by so-called progress?

Fortunately more and more protests are raised by those realizing the danger. Often their protests are deliberately ignored because they are uncomfortable, need financial support, and do not promise material profit. But we have no right to exploit nature unscrupulously. On the contrary, it is our duty to find new refuges for animals and plants to preserve them for the future. In the case of orchids it is difficult but not impossible to cultivate endangered species. This is true of European species, of which many are in danger, at least locally or regionally.

Botanic gardens should offer a home for endangered plants. Co-operation with the local office for environmental protection and other conservation and land management authorities is advisable. Also we should exchange information on the results and experiences of the culture of European orchids, a group that is so much more difficult to grow than most other plants. Their cultural needs differ considerably from one species to another and it is not possible in one paper to give the details for every one. So only general hints can be given.

No botanic garden should aim to build up a collection of too many different species, but should rather try to keep the local species, because the natural climatic conditions for such plants should be the easiest to maintain.

In pot culture, orchids should be grown in rather large pots. I would advise six to ten plants together in a 16–18 cm pot. Good drainage at the bottom is most important; the compost should be as similar to soil in the wild habitat as possible and should be covered with pine needles to prevent the growth of moss or a crust forming on the surface. A little humus should be added as this has

207

a favourable effect on fungal growth. Cultivating orchids in pots enables the plants to be moved easily so that differing seasonal demands can be taken care of and flowering specimens can be shown at exhibitions. The location of the pots depends on the species: pots of frost-hardy orchids can be placed in the open ground, those of more tender species in a closed frame where they can be protected from frost, too much rain and too strong sunlight. Many species are sensitive to late spring frosts.

For Mediterranean orchids the best place is in a glasshouse. My ideal house for European orchids would be low and built into the ground (such a house is called a 'pit' in England). The pots stand on both sides in natural soil at ground level. Plenty of ventilation is important, both from the roof and from the sides. Best of all is a house in which the sides can be moved away altogether in good weather, so that the air can move freely around the plants. Automatic mist equipment should be installed and for those species with leaves in winter some additional lighting is required. The roofs should be well shaded in summer. Heating should keep the house just above frost level at night, but during the day, with growing light intensity, temperatures may rise to about 20°C. A large difference between day and night temperatures is desirable, and is possibly absolutely necessary.

Most critical is the air humidity. Measurement in the field showed the following results on a hot sunny day:

Height above soil level	Relative humidity
1.50 m	30%
0.70 m	60% (near the flowers)
0.10 m	90% (near the leaves)

Add to this a heavy dew formation in the morning, much of which is taken up by the leaves, and the plant is by no means as dry as one might imagine! In the natural habitat long droughts during the vegetative period cause earlier maturing and dying down of the leaves; far fewer seeds are formed in such years. The tubers or pseudo-bulbs are smaller than normal and in the following year the plants often fail to flower. This should be borne in mind when growing orchids in cultivation, especially in the spring when humidity can drop rapidly on sunny days. Sun is important, but adequate humidity is indispensable.

Strong ventilation also seems to be absolutely necessary. On frosty days, when no ventilation was possible, leaves and rosettes frequently rotted. A ventilator at plant level can bring some movement of air, but it is only a makeshift. There are now, however, other technical means enabling warm air to be blown into the house.

Orchids appreciate weak fertilizer during the vegetative period, contrary to the opinions expressed in books on the subject. We have used different fertilizers in our garden, alternating with manure, with good results. As soon as the leaves begin to yellow, addition of fertilizer should be stopped.

Various pests can be controlled by some insecticides, e.g. 'Metasystox', without damage. Great care should be taken, however, not to use too many insecticides on orchids. The resistance developed by the pests stimulates the use of new insecticides and it seems far better to prevent damage by keeping the plants in good conditions. Snails can do terrible damage to orchids in pots, where they find plenty of hiding-places, particularly between the pots; 'Slugit' and similar compounds should always be to hand. Birds can cause damage quite unexpectedly when they dig for worms inside the frames or low glasshouses; the use of nets or of fine wire netting can prevent this.

Propagation should be attempted. The method of asymbiotic sowing still proves unproductive, but the old method of sowing seed around an established plant can be more successful. Here a loose cover of humus-forming material on the soil is a great advantage to prevent drying out of the soil and stimulate growth of fungi. We noticed the first green leaves on seedlings of *Ophrys sphegodes* three months after sowing and the first of the young plants flowered in the third year.

For raising orchids *in vitro*, there are several alternative 'recipes' for the culture medium. From my own experience, I have had good results with the following:

> 10 g Fructose
> 10 g Agar
> 3 g Peptone
> 3 g Activated charcoal
> 2 g Yeast extract
> 10 drops Polybion
> Distilled water

I achieved success with *Orchis, Ophrys* and *Barlia* on this medium with or without minerals, often within four to eight weeks after sowing. Within 10 to 12 months the seedlings reached a size of 2–3 cm and could be thinned out. After that, however, very great losses occurred, sometimes amounting to 100 per cent.

Vegetative propagation can be undertaken at two stages of growth. If a plant such as *Orchis morio* (in October or November) has produced enough leaves and roots and is growing well, one can cut off the tuber or pseudo-bulb under the roots. After some weeks this will form one, or sometimes two, new shoots. The parent plant grows in the meantime without its tuber or pseudo-bulb and will form a new one for the following year. The second opportunity is during or soon after flowering, as long as the leaves are still green. At this instant the newly formed bulbous rootstock for the next year is removed from the plant and planted by itself. On the old plant remains an 'eye', which will form another rootstock, sometimes even several of them. They remain quite small during the first year.

In this way no great quantity of new plants is produced, but it is possible to obtain material for meristem propagation without having to destroy the parent plant. No optimum conditions have yet been found for this method of propagation and the greatest difficulty remains planting out into the open ground. Many things we have yet to learn, lots of difficulties have still to be solved—a rewarding subject for a scientific dissertation!

The cultural methods described for European orchids can certainly have satisfactory results, but the best way to conserve them is still to care for them in their natural environment and maintain their habitats. The most important measure is the control of competition from other plants. Some trees and shrubs are valuable to keep off desiccating winds and too much sun. But if they get out of hand, as for examples Sloes (*Prunus spinosa*) frequently do, the orchids are at a great disadvantage and the shrubs should be removed. Also a thick growth of grass can suffocate orchids. Such grass should be mown, but only when the orchid leaves have disappeared completely. A little of the grass can be used as mulch, but large amounts are better removed to the compost heap. The well-rotted material can be returned later and spread over the orchid site. Mowing should not take place too late in the year, in order to allow grasses and other plants to maintain growth before the winter.

Quite often a poor seed set is observed in a natural orchid stand. The weather might be cool and moist at flowering time, and the necessary insects fail to appear. Here pollination by hand can help a great deal, although a lot of patience is needed for this work, especially for orchids with small flowers like *Gymnadenia* and *Spiranthes*. One single seed pod contains a large number of seeds, but many losses can be expected from wild animals, pests and so forth. Therefore it is better to pollinate as many flowers as possible. Also the sowing of the seed should not be left to the wind, but the seed from the pods collected in a paper bag (not plastic!) shortly before they open. They can complete maturation in the bag and one can decide where to sow them around the parent plants. This effort will show results possibly after three years.

In conclusion I would like to say that one can learn best about the cultivation of the European orchids by observing them closely in their natural habitat all the year round, especially since many of the principles for cultivating other plants do not apply to this group. I would be happy if these observations could help even a little to save some of our most endangered European orchids.

Trying to Conserve the Rare and Endangered *Degenia*

VINKO STRGAR

University of Ljubljana, Yugoslavia

Degenia velebitica is one of the rarest and most threatened plants in Europe. In this paper I shall show how it is thriving in nature and how we are endeavouring to preserve it in the Ljubljana Botanical Garden.

Degenia velebitica (Degen) Hayek (Family Brassicaceae) is a perennial, caespitose, silver-grey herb with non-flowering rosettes. Its stems are up to 10 cm high; the leaves are linear-lanceolate, 10–15 mm long and 2–4 mm broad. The flowers are yellow, 10–12 mm broad. The fruit is an ellipsoidal silicula with inflated valves; there are 2 seeds in each loculus.

The genus most akin to *Degenia*, which itself only contains a single species, is *Lesquerella* from North America. In nature, the *Degenia* is known from just two localities in the Velebit mountain range in the Dinaric Alps of Yugoslavia. One locality is in the southern Velebit in the region of Šugarska Duliba with four small sites at 1200–1400 m above sea-level (Degen, 1909, 1937; Rossi, 1911; Hayek, 1910); the second locality is situated in the northern Velebit on the mountain of Solila about 1150–1200 m above sea-level (Horvat, 1953). The *Degenia* grows on the most wind-exposed, small, sunny, dry carboniferous screes, very rarely on fragmented screes and among loose rocks. In Šugarska Duliba one site is on a well developed scree covering an area of about 300 sq. m where 63 plants were found flowering in 1975, while in the remaining three sites just single plants are found on scree fragments and among loose rocks. The situation is a little better in the northern Velebit.

Phytosociological research on the *Degenia* habitats in the southern Velebit (Horvat, 1930; Horvat *et al.*, 1974) and in the northern Velebit has shown that the *Degenia* mostly grows and reaches optimal development in the scree plant association of *Bunio-Iberetum pruitii* (as *Bunieto-Iberetum carnosae* in Horvat, 1931), which can be found on dry limestone screes of several Balkan mountains.

There are many reasons why we are most interested in this plant, particularly from the botanical, horticultural, nature conservation and protective point of view. Firstly, as a monotypic relict genus in Yugoslavia, the *Degenia* is botanically an extraordinarily interesting and rare phenomenon. Secondly, being a rare plant, the *Degenia* is attractive to the discerning plantsman and gardener; a plant so very beautiful in blossom and even more so in fruit could be suitable for wider horticultural purposes. In the third place, in our opinion at the moment the most important, the *Degenia* is a rare and endangered plant worthy of our endeavours to protect and preserve it. To deal more fully with this last point we should first look at what threats there are to the *Degenia* and how we can save

211

it. The greatest and in its final stages the most fatal enemy of the *Degenia* so far identified by previous authors (Horvat, 1930) and by ourselves, is the gradual natural and for the *Degenia* unfavourable change in the habitat. The small screes on which the plants occur are under low rocky outcrops yielding ever smaller quantities of mineral substrate. The screes gradually stabilize, become enriched with humus and overgrown by meadow plant communities. On former habitats of scree plant associations, instead of *Degenia velebitica, Iberis pruitii, Bunium alpinum* and other species, meadow elements such as *Helianthemum balcanicum*, and later, *Sesleria juncifolia, Carex humilis, Arctostaphylos uva-ursi, Satureja subspicata, S. montana* and other species, keep on appearing and completely ousting the *Degenia*, leaving the only habitat open to the species as the small and fragmented, bare screes which are perpetually preserved because of the action of strong winds. The second danger for the *Degenia*, at the moment greater than the first threat, though when viewed on a longer time scale we hope is less important, is caused by herbalists and those botanists, gardeners, amateurs and others who, in spite of the *Degenia* being protected by law, still pick the plants and remove the seeds in order to grow it. Such gathering, especially by herbalists who gather the most beautiful fruit-producing specimens, seriously endangers the species since it is already rare in nature. It was Horvat in 1930 who stated that in the southern Velebit in all four sites there were likely to be fewer than a hundred fully developed specimens of the *Degenia*; our own studies confirm this statement, as does Kušan (1963).

How are we planning to protect the *Degenia*? The law, formally protecting it from gathering, is truly a needed, praiseworthy and important tool. However,

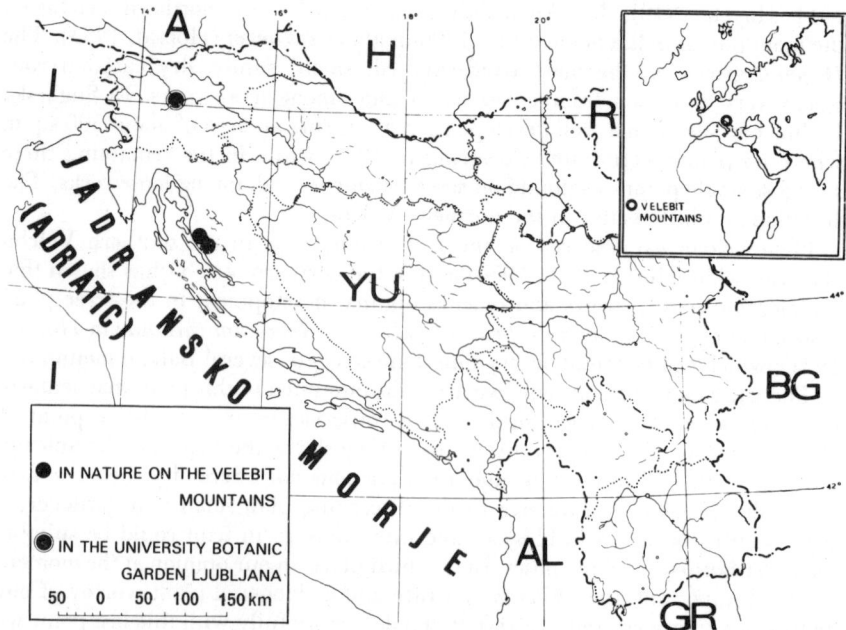

FIGURE 1. Geographical Distribution of *Degenia velebitica*

the law is powerless when dealing with the overgrowing of the screes; also it cannot prevent the picking of the *Degenia* in the natural habitats as it cannot be efficacious in the solitary and remote Velebit mountains, unreachable by the arm of justice that would be able to enforce the law and actually protect the *Degenia*. Thus more is needed to actively safeguard the *Degenia*, which means controlling the overgrowing of its habitats, protecting the plants from collectors and, in the last resort, restoring the *Degenia* on the screes with plants and seeds from cultivation.

Active consideration on how to prevent the *Degenia* habitats from becoming overgrown has not been elaborated yet. The overgrowing of the screes is a process lasting centuries and even longer; it is also likely to take centuries before other plants take over all the *Degenia* habitats by natural overgrowth—thus, in our opinion, there need be no hurried action, realizing that in the southern Velebit there is still approximately the same number of mature plants as there were 50 years ago.

How to protect the *Degenia* directly and effectively from herbalists and plant growers is difficult to visualise. An indirect way may be easier: we should study the prospects of growing the *Degenia* and giving individual plants from cultivation to those who are interested in them. This idea has other advantages as it involves an additional practical aim. In numerous experts' opinion, the *Degenia* is a beautiful decorative plant. Mathew (1975) thought much of this plant, as the following sentence indicates: "This inhabitant of the Velebit Mountains has great value as a garden plant, for it has intensely silver foliage serving as a foil to the flowers in summer, and then for the large bladder-like seed vessels". Such a point of view on the prospects for these species in horticulture and the need to protect it in the wild has induced us in Ljubljana Botanical Garden to start research along the lines of introducing native plants into cultivation, undertaking more detailed study and making experiments on a wider scale to find out how to grow this rare and interesting plant. It would be appropriate here to mention some of the existing experiments for growing a small number of *Degenia* plants carried out by botanists, gardeners and amateurs. The first experimental culture, mainly for taxonomic purposes, was started in the botanical garden of Budapest immediately after the discovery of the *Degenia* around 1908 (Degen, 1937). The first (unsuccessful) trial in the botanic garden at Ljubljana took place in 1937. Twenty years later, the plant was again brought to Ljubljana where we succeeded in growing it and it is now being regularly reproduced by seed and from cuttings, so that during the last 20 years, in Ljubljana at least, a few plants have been blossoming and producing fruits (Strgar, 1971). The plant was also successfully grown in other botanic gardens (Zagreb, Botanical Garden of Velebit). Schwarz (1974) reports on growing a *Degenia* in the alpine house at Jena. Several plants grew successfully also in Mrs Dryden's alpine house in England (Gorer, 1975) and in the botanic garden at Copenhagen (Olsen, pers. comm.) and probably elsewhere.

Finally, a few words on our present, wider research on the prospects of growing the *Degenia*: apart from the work on germination and other aspects which we started in 1973, our work consists in gathering and studying the already known data on the *Degenia* in the wild and in cultivation. In addition, it covers ecological, phytosociological, phenological and other occasional observations and measurements in nature, and experiments on propagating and growing plants in improvised habitats in the garden.

In the wild and in cultivation our attention is being given predominantly to conditions of the soil, weather and climate, the greatest effort being put into phenological observations consisting, among others, in counting sprouts, flower buds, blossoms, fruits and seeds of a certain number of chosen plants living in the wild and in cultivation, and correlating this data. We have now about 700 plants in cultivation; 272 of these blossomed and produced fruits in 1978. There is now quite a large amount of additional data which cannot be discussed here, but it is important to say that the results of the experiments show that, considering the plant's main peculiarities, at least under the conditions prevailing in Ljubljana, the *Degenia* can be grown without too many problems, very probably even in the open.

REFERENCES

DEGEN, A. (1909). Ueber die Entdeckung eines Vertreters des Gattung *Lesquerella* im Velebitgebirge. *Magyar Bot. Lapok.* **8**: 3–29.
——(1937). *Flora Velebitica, Vol.2.* Budapest. Pp. 191–198.
GORER, R. (1975). *Degenia velebitica* (Degen) Hayek P. C. *Quart. Bull. Alp. Gard. Soc.* **43**: 317.
HAYEK, A. (1910). Die systematische Stellung von *Lesquerella velebitica* Degen. *Österr. Bot. Zeitschr.* **60(3)**: 89–93.
HORVAT, I. (1930). Vegetacijske studije o hrvatskim planinama. I. Zadruge na planinskim goletima. *Rad Jugoslav. Acad. Znan. Umjet.* **238**.
——(1931). Vegetacijske studije o hrvatskim planinama. II. Zadruge na planinskim stijenama i točilima. *Rad Jugoslav. Acad. Znan. Umjet.*
——(1953). Prilog poznavanju raširenja nekih planinskih biljaka u jogoistočnoj Evropi. *Godišnjak Biol. Inst. Sarajevu* **5(1–2)**: 199–217.
——, V. Glavač & H. Ellenberg (1974). *Vegetation Südosteuropas.* Gustav Fischer Verlag, Stuttgart. 768 pp.
KUŠAN, F. (1963). Über die Lebensverhältnisse der endemischen Art *Degenia velebitica* (Deg.) Hay. auf dem Velebit in Kroatien. *Informationes (Edicio Periodica Horti Botanici facultatis pharmaceuticae Universitatis Zagrebensis) Jugoslavia* 2: 21-26.
MATHEW, B. (1975). Some interesting plants at the Shows, 1975. *Quart. Bull. Alp. Gard. Soc.* **43**: 291–306.
ROSSI, L. (1911). U. Šugarskoj Dulibi. *Glasn. Hrv. Prirodosl. Društva* **23**: 1–47.
SCHWARZ, O. (1974). The alpine house at Jena. *Quart. Bull. Alp. Gard. Soc.* **42**: 188–189.
STRGAR, V. (1977). *Degenia velebitica. Quart. Bull. Alp. Gard. Soc.* **45**: 4–5 (Translation of part of an article in *Proteus, Ljubljana* **33**: 401–404 (1971).)
——in T. Wraber (1971). Na obisku v Ljubljanskem botaničnem vrtu. *Proteus, Ljubljana* **33**: 9–10, 399–405.

The Rôle of Collections of Wild Species as Seed Orchards for the Cultivation of Unusual Plants

P. C. de JONG

University Botanic Garden, Utrecht, Netherlands

Exotic woody plants are an important element of the Dutch landscape, notably in public grounds and gardens. Trees like the Horse Chestnut, Service Berry (*Amelanchier*), Scots Pine and False Acacia became naturalized some hundred years ago and enriched our local flora, which contains, unfortunately, only a modest number of woody plants.

Dutch nurserymen have an excellent reputation for the cultivation of exotic plants, for instance with conifers, rhododendrons, Japanese Maples and roses; more than half of their production is exported to other West European countries. A visitor to any of the small nurseries would find that they held only a few mother (or stock) plants and very rarely are such plants kept just for seed production. In the case of cultivated varieties vegetative propagation is essential, but this method of propagation is also quite commonly used for true species.

The acquisition of seeds to produce stock of wild species depends on rather fluctuating supplies from abroad and more often seeds are collected in public grounds and botanic gardens. Frequently these seeds are collected from single plants or poor specimens. The progeny resulting from these sowings is often disappointing and more and more nurserymen give preference to vegetative propagation for the following reasons:

1. *The true identity of the material* can be guaranteed better. The selected plants are of known performance and so will give satisfaction to the customers.
2. *The quality* generally is better, both as a consequence of selection and because vegetative propagation is used.
3. *The cost of production* will be lower, through faster development of the plants, and there will be less effort and cost in obtaining propagating material.
4. *The continuity* of production will be guaranteed by the more regular supply of cuttings and (if necessary) of rootstocks.

This cloning of wild species started a long time ago with for example conifers and *Rhododendron* species, but also with species of *Acer*, *Populus* and *Quercus*. *Quercus frainetto* from central Europe is often used in Holland as a street tree; all these trees originate from a single plant. All *Acer saccharinum* plantings go back to four clones. All individuals of *Acer cissifolium* have been grown from one female plant.

Botanic gardens often lack the facilities to produce most of their own stock of woody plants, and therefore purchase many rare plants direct from specialized nurseries. Thus several of these 'clonal species' may also be found in botanic gardens of western Europe. Future computer printouts may give the impression that the species concerned are not rare in cultivation, but in many cases all gardens have only a part of one and the same plant. It will not be an easy matter to alter this unfortunate development. Nurserymen themselves have no need to change their methods of cultivation. The customers are mainly interested in having the material true to species, and they are rarely, if ever, interested in the origin of the material and the method of propagation. Growers will only become interested when there are continuous supplies of seeds available, which can give rise to plants of the same quality, and for the same cost, as vegetatively propagated stocks. Experience in forestry with important exotic trees such as Douglas Fir and Scots Pine has shown that the import of seeds is a difficult matter. Only seeds from certain natural sources will grow well under Dutch climatic conditions. So it is easy to understand why the import of seeds of rare and unusual plants may frequently lead to disappointing results.

What kind of solutions should be proposed? In our opinion we should start seed orchards in which vegetatively propagated specimens of well acclimatized trees and shrubs are planted in groups that are well isolated from related species. The same may be undertaken with new introductions from natural sources. The offspring of such groups have to be tested before they are distributed. Therefore a committee consisting of both botanists and horticulturists has to be involved in the evaluation of these offspring.

Such a committee already exists in the Netherlands for the selection of ornamentals. It was formed by the N.A.K. (Netherlands General Inspection Service) which has sections for vegetables, cut-flowers, fruit, seed potatoes and ornamentals. Some inspections are free, others are obligatory; for instance all fruit trees have to be inspected at the nurseries and labelled before sale. In the case of ornamentals, inspection and labelling is free, but an increasing number of nurserymen have their plants inspected because labelled plants sell better. So there is already an official organization in existence in the Netherlands, which is able to undertake this task.

In the Netherlands the Institute for Forestry and Landscaping 'De Dorschkamp' produces an inventory of all large and old exotic plants growing throughout the country. This also includes any well acclimatized unusual plants. The documentation is to be computerized. Several fine specimens of species rare in cultivation have been found and propagation of these plants is starting now. In the Gimborn Arboretum, a satellite of the Utrecht Botanic Gardens, we hope over the coming years to plant groups of these plants isolated from related species. The co-operation of other institutes and botanic gardens, both in the Netherlands and abroad, is necessary for real success in this project.

A successful example of this method of production is demonstrated by the results achieved with the Paperbark Maple (*Acer griseum*). The plants of this species in general cultivation are normally rather slow-growing, and rarely produce fruits which give rise to good offspring. At Hergest Croft in Herefordshire, England, there are two very large, well acclimatized trees grown from seeds collected by Wilson in China. The seedling offspring of these specimens are very vigorous and produced shoots over 1.25 m long last summer. This is a much better result than may be achieved by grafting. So this example shows that a good selection of mother trees can make it possible to surpass results which are normally only obtained by vegetative propagation.

CONCLUSION

In reconsidering the valuable material that was introduced from earlier expeditions abroad, we find that specimens are often scattered over Western Europe. In many instances we have lost trace of introductions, because we do not now know precisely where they are and which specimens have survived. In analysing the existing valuable material, such as is achieved by the survey in Holland, we can reconstitute part of what we believed to be lost. Through this analysis we can bring together valuable material and we can use it as the base from which to start.

This original material can give us the seedlings or cuttings to build up new collections. It has been well acclimatized under our climatic conditions and any less desirable genetic material has been eliminated in the process. The offspring of such collections can give a very valuable contribution to both horticulture and conservation.

Creating Specialized Habitats in a Garden

MARK SMITH

University of Bristol Botanic Garden, England

I will describe attempts made at the University of Bristol Botanic Garden to make and maintain a small artificial sand dune and *Sphagnum* bog, and will comment on some plants that have succeeded and others that have failed to grow well.

SAND DUNE

The sand dune is made of sand dredged from the Bristol Channel. It measures 8 metres by 3 metres, including the surrounding stone walls that are 50 cm thick. It has two levels of sand, 75 cm and 150 cm deep, separated by a wall running down the middle. Our sand dune is in effect a raised bed having two steps, facing towards the southwest. The functions of the thick walls are to stop the sand blowing away, to provide paths along their flat tops, and to reduce the rate of water loss. In front of the sand dune the ground has been dug out and replaced by a row of eight square concrete pools with paving slabs as surrounds. These were designed to simulate a dune slack and salt marsh.

The sand dune was built in March 1977 and planted and sown in the following two months. I had expected that during the first summer, before the plants had grown large, the sand might be blown off the surface; as a precaution the bed was not filled quite to the top with sand and boards were put across as wind-breaks. In fact very little sand was lost, and by the second summer a hard crust had developed so that there was little movement of sand even in bare areas. There is provision to water the areas with a sprayline, but apart from the period just after planting this has hardly been necessary during the last two wet summers. We have not resown annual species, but are relying on self-seeding.

Nearly all the plants grown were obtained as offsets or seed from the sand dunes at Berrow in Somerset and Braunton in North Devon. I will list some of them, under four groups.

1. Species that have grown well in the 'sand dune', but which also grow well in ordinary garden soil at Bristol, where the soil is a heavy, lime-rich loam: *Hippophaë rhamnoides, Salix repens, Crambe maritima, Raphanus maritimus, Plantago coronopus, Trifolium arvense, Euphorbia portlandica, Glaucium flavum, Erodium cicutarium.*

2. Species that have grown well in the sand, but which we have not tried to

219

grow in ordinary soil: *Elymus arenarius, Carex arenaria, Agropyron juncei-forme, Vulpia membranacea, Cakile maritima, Arabis brownii, Viola canina, V. tricolor* subsp. *curtisii.*

3. Species that grow well in the sand, but not in ordinary soil: *Ammophila arenaria, Phleum arenarium, Eryngium maritimum, Erodium maritimum, Euphorbia paralias, Centaurium scilloides, Matthiola sinuata.*

4. Species that have not grown well in the sand: *Calystegia soldanella*, which was difficult to establish (from roots) but is now growing and has passed through one winter. Although abundant seed of *Atriplex laciniata* and *Salsola kali* was produced in the first year, only a few seedlings appeared during the second year, and the form of the plants is less succulent and more diffuse than those growing by the sea. Next year we will add salt to see if this improves growth.

In the 'dune slack' pools *Equisetum variegatum* and *Epipactis palustris* have grown well. We have not yet tried any salt marsh plants.

SPHAGNUM BOG

A concrete-lined garden pond, 15 metres long and from 1 to 2 metres across, was converted into a bog by lining it with polythene and filling with peat. A concrete border was made around part of the peat bed, sloping inwards; its purpose was to catch additional rain-water, and it was painted with a silicone resin to prevent the uptake of lime. Half the surface of the peat was planted with *Sphagnum* in October 1975; this died in the following summer when Bristol had a 60-day drought. More *Sphagnum* was planted in late 1976, and it has grown well although extra water has had to be supplied frequently during the summer. The moss was collected at peat deposits near Glastonbury, Somerset, by courtesy of the Eclipse Peat Company. It was initially composed of almo st equal parts of *Sphagnum plumosum* and *S. tenellum*; after two years most of the former has died and the latter has spread to take its place.

Some of the plants we have put into this bog have failed to grow well and in some cases have died out. At one end, forming a group of tall plants, *Myrica gale, Osmunda regalis, Carex paniculata* and *Peucedanum palustre* have done well; in contrast *Eriophorum angustifolium* and *Cladium mariscus* have grown slowly. In the part covered by *Sphagnum*, *Anagallis tenella* has spread rapidly; *Pinguicula grandiflora* has grown well and increased vegetatively, but few seedlings have appeared; *Drosera rotundifolia* has persisted but not increased; *D. anglica* has decreased; *Lycopodium inundatum* has died out. In wet areas we have tried and failed to maintain *Hypericum elodes* and *Pilularia globulifera*.

Maintenance has been difficult in dry weather. Since our tap-water is alkaline the bog has had to be watered by siphoning from greenhouse rain-water tanks 100 metres away. Our tanks have twice been emptied after dry spells of three weeks and we have had to water with tap-water. The lime content of the peat may also have been raised by leaf-fall off nearby trees and by dust. When there has been enough rain to refill our tanks we have tried to flush lime out by overflowing the bog with rain-water. When the bog was made we used sedge peat from Somerset, with an initial pH of 5–6: we plan to replace this with Irish moss peat of pH 3–4.

While encouraged by having kept the *Sphagnum* in good condition for two years, I consider our success rate with other plants has been poor. I attribute this to difficulty in supplying sufficient water, damage by birds and temporary increases in lime content.

List of Delegates

Dr C. D. ADAMS, Department of Biological Sciences, University of the West Indies, St Augustine, Trinidad.

Mrs A. ALA, Botanical Institute of Iran, PO Box 8-6096, Tehran, Iran.

Mr P. ALANKO, Botanical Gardens of the University, Unioninkatu 44, 00170 Helsinki 17, Finland.

Mr D. ALDRIDGE, Countryside Commission for Scotland, Battleby, Redgorton, Perth, United Kingdom.

Dr A. AMARAL, Botanisches Institut und Botanischer Garten der Universität Wien, Rennweg 14, 1030 Vienna, Austria.

Herr J. APEL, Institut für Allgemeine Botanik und Botanischer Garten, Hesten 10, 2000 Hamburg 52, Federal Republic of Germany.

Mr F. ARNKLIT, Botanisk Have, Københavns Universitet, Øster Farimagsgade 2B, 1353 Copenhagen, Denmark.

*Dr L. V. ASIESHVILI, Central Botanical Garden, Academy of Sciences of the Georgian SSR, PO 5, 380005 Tbilisi, Georgian SSR, USSR.

Dr M. AVISHAI, Botanical Gardens, The Hebrew University, Jerusalem 91000, Israel.

Dr G. W. M. BARENDSE, Department of Botany, University of Nijmegen, Toernooiveld, Nijmegen, Netherlands.

Herr H. BECELA, Botanischer Garten der J. W. Goethe Universität, Siesmayerstr. 72, 6000 Frankfurt-am-Main, Federal Republic of Germany.

Dr D. BELLAMY, Department of Botany, University of Durham, South Road, Durham, United Kingdom.

Mr R. I. BEYER, Royal Botanic Gardens, Kew, Richmond, Surrey, United Kingdom.

Mr L. BISSET, Royal Botanic Garden, Edinburgh EH3 5LR, United Kingdom.

Dr J. van BORSSUM WAALKES, Hortus de Wolf, State University of Groningen, Kerklaan 30, Haren (Gr.), Netherlands.

Mr A. BRADY, National Botanic Gardens, Glasnevin, Dublin 9, Eire.

Dr D. BRAMWELL, Jardín Botánico 'Viera y Clavijo', Apto 14, Tafira Alta, Las Palmas de Gran Canaria, Islas Canarias, Spain.

Professor J. P. M. BRENAN, Royal Botanic Gardens, Kew, Richmond, Surrey, United Kingdom.

Mrs M. BRIGGS, Botanical Society of the British Isles, White Cottage, Slinfold, Horsham, Sussex, United Kingdom.

Mr P. W. BRISTOL, The Holden Arboretum, 9500 Sperry Road, Mentor, Ohio 44060, U.S.A.

Mr B. F. BRUINSMA, University Botanic Garden, Nonnensteeg 3, Leiden, Netherlands.

Mr P. CHAI, Ibu Pejabat Jabatan Hutan (Forest Department Headquarters), Kuching, Sarawak, Malaysia.

223

Mr M. B. CHAICHI, Botanical Institute of Iran, PO Box 8-6096, Tehran, Iran.

Mr M. CHAOUAT, Mt Scopus Botanical Garden, The Hebrew University of Jerusalem, Jerusalem, Israel.

Mr E. B. CHEW, Zoological Society of San Diego, PO Box 551, San Diego, California 92112, USA.

Mrs G. CROMPTON, University Botanic Garden, Cambridge CB2 1JF, United Kingdom.

Mr D. L. CUNNINGHAM, David L. Cunningham Inc., 215 Riverside Drive, Newport Beach, California 92663, USA.

Mr E. W. CURTIS, Botanic Gardens, Glasgow G12 0UE, United Kingdom.

Monsieur S. van DIEVOET, Administration de la Donation Royale, Arboretum Géographique de Tervuren, Avenue du Derby 55, Boite 6, 1050 Brussels, Belgium.

Mr D. DONALD, University Botanic Garden, Cambridge CB2 1JF, United Kingdom.

Dr J. F. DURAND, The John F. Kennedy Park, New Ross, Co. Wexford, Eire.

Sr M. G. EMDE, Hortus Botanicus 'Marimurtra', Fundación Carlos Faust, Estación Internacional de Biologia Mediterránea, Blanes, Gerona, Spain.

Dr H. ERN, Botanischer Garten und Botanisches Museum Berlin-Dahlem, Königin-Luise-Strasse 6–8, 1000 Berlin 33, Federal Republic of Germany.

Mr C. M. ERSKINE, Royal Botanic Gardens, Kew, Richmond, Surrey, United Kingdom.

Miss L. FARRELL, Nature Conservancy Council, PO Box 6, Godwin House, George Street, Huntingdon, United Kingdom.

Mr H. J. FLIEGNER, Royal Botanic Gardens, Kew, Richmond, Surrey, United Kingdom.

Dr L. GODICL, University of Maribor, Pedagogical Academy of Maribor, Koroška c, 62000 Maribor, Yugoslavia.

Professor C. GÓMEZ-CAMPO, Escuela Tecnica Superior de Ingenieros Agronomos, Universidad Politécnica, Madrid 3, Spain.

*Dr E. E. GOGINA, Herbarium, Main Botanic Garden, Academy of Sciences of the USSR, Botaniceskaya 4, Moscow 127276, USSR.

Mr P. S. GREEN, Royal Botanic Gardens, Kew, Richmond, Surrey, United Kingdom.

Dr A. S. GUERRA, Apto 60, La Laguna, Tenerife, Islas Canarias, Spain.

Mr B. HALLIWELL, Royal Botanic Gardens, Kew, Richmond, Surrey, United Kingdom.

Mrs P. HARITONIDOU, 21 Tayetou Street, Psichico, Athens, Greece.

Mr A. R. HASSAN KING, Estate Department, Njala University College, Private Mail Bag, Freetown, Sierra Leone.

Mr S. J. HENCHIE, Royal Botanic Gardens, Kew, Richmond, Surrey, United Kingdom.

Monsieur J. IFF, Conservatoire et Jardins Botaniques, Case postale 60, 1292 Chambesy, Geneva, Switzerland.

Dr S. K. JAIN, Botanical Survey of India, Botanic Garden, Howrah-711103, West Bengal, India.

Mrs Z. JAMZAD, Botanical Institute of Iran, PO Box 8-6096, Tehran, Iran.

Dr D. W. JEFFREY, Trinity College Botanic Garden, Botany School, Trinity

College, Dublin 2, Eire.

Professor K. JONES, Royal Botanic Gardens, Kew, Richmond, Surrey, United Kingdom.

Dr P. C. de JONG, Botanische Tuinen, Rijksuniversiteit Utrecht, Heidelberglaan 2, Utrecht 2506, Netherlands.

Mr C. P. KELLY, The John F. Kennedy Park, New Ross, Co. Wexford, Eire.

Mr E. E. KEMP, Dundee University Botanic Garden, 516 Perth Road, Dundee, United Kingdom.

Dr A. KHALIGHY, Department of Horticulture, College of Agriculture, Karaj, Iran.

Mr B. KIRCHNER, Botanischer Garten der Ruhr-Universität, 4630 Bochum, Postfach 102148, Federal Republic of Germany.

Dr K. KOZAK, Universytet Marii Curie-Skłodowskiej, Ogród Botaniczny, ul. Sławinkoska 3, 20-810 Lublin, Poland.

Mr H. KRAFT, Botanischer Garten und Botanisches Museum Berlin-Dahlem, Königin-Luise-Strasse 6-8, 1000 Berlin 33, Federal Republic of Germany.

Mr H. KUHBIER, Übersee-Museum, Bahnhofsplatz 13, 2800 Bremen 1, Federal Republic of Germany.

Ir. E. LAMMENS, Jardin Botanique National de Belgique, Domaine de Bouchout, 1860 Meise, Belgium.

Mr C. R. LANCASTER, The Hillier Arboretum, Jermyns Lane, Ampfield, Nr Romsey, Hampshire, United Kingdom.

Dr T. LASSER, Jardin Botánico de la Universidad Central, PO Box 2156, Caracas, Venezuela.

Professor T. B. LEE, Kwanak Arboretum, College of Agriculture, Seoul National University, Kyunggido, Suwon, Korea 170.

Mr J.-Y. LESOUËF, Conservatoire Botanique du Stangalarc'h, 29200 Brest, France.

Mr G. LL. LUCAS, Royal Botanic Gardens, Kew, Richmond, Surrey, United Kingdom.

Mr MARCZEWSKI, Botanical Garden, Polish Academy of Sciences, 02-973 Warsaw, 34 skr. 84, Powsin, Poland.

Mr E. MARKER, University of Oslo, Botanical Garden, Trondheimsvn. 23B, Oslo 5, Norway.

Mr P. J. MAUDSLEY, Department of Botany, University of Durham, South Road, Durham, England.

Mr W. F. MEARNS, Wollongong Botanic Gardens, Keiraville, New South Wales 2500, Australia.

Mr R. J. MITCHELL, Botanic Garden, The University of St Andrews, Fife KY16 8RT, United Kingdom.

Professor G. MOGGI, Orto Botanico dell'Università, Via P. A. Micheli 3, 50121 Florence, Italy.

Dr E. F. MOLCHANOV, State Nikita Botanical Garden, 334267 Yalta, Crimea, USSR.

Dr B. A. MOLSKI, Botanical Garden, Polish Academy of Sciences, 02-973 Warsaw, 34 skr. 84, Powsin, Poland.

Mr K. D. MORGENSTERN, 50 Buttermere Court, Boundary Road, London NW8 6NS, United Kingdom.

Mr J. K. MUIR, City of Liverpool Botanic Garden, The Mansion House, Calderstones Park, Liverpool, United Kingdom.

Dr O. E. G. NILSSON, Botanical Garden, Villavägen 8, S-75236 Uppsala, Sweden.

Monsieur L. OLIVIER, Parc National de Port-Cros, 50 Avenue Gambetta, 83400 Hyères, France.

Mr P. ORRISS, University Botanic Garden, Cambridge CB2 1JF, United Kingdom.

Mr K. OTTEN, Plantentuin der Rijksuniversiteit, K. L. Ledeganckstraat 35, 9000 Gent, Belgium.

Dr F. H. PERRING, Monks Wood Experimental Station, Institute of Terrestrial Ecology, Abbots Ripton, Huntingdon, United Kingdom.

Dr H.-H. POPPENDIECK, Institut für Allgemeine Botanik und Botanischer Garten, Hesten 10, 2000 Hamburg 52, Federal Republic of Germany.

Dr G. T. PRANCE, The New York Botanical Garden, Bronx Park, Bronx, New York 10458, USA.

Mr J. R. PRATT, Duffryn Gardens, St Nicholas, Cardiff CF5 6SU, United Kingdom.

Dr V. PULEVIĆ, Republički zavod za zaštitu prirode, PO Box 2, 81001 Titograd, Yugoslavia.

Mr D. RADLEY, Department of Plant Biology, University of Birmingham Botanic Garden, 56 Edgbaston Park Road, Birmingham, United Kingdom.

Frau I. von RAMIN, Palmengarten, Siesmayerstrasse 61, 6000 Frankfurt-am-Main 1, Federal Republic of Germany.

Dr D. A. RATCLIFFE, Nature Conservancy Council, 19/20 Belgrave Square, London SW1X 8PY, United Kingdom.

Mr A. M. REGUEIRO, Real Jardín Botánico, Claudio Moyano 1, Madrid 7, Spain.

Mr W. RICHTER, Neuer Botanischer Garten der Universität Göttingen, Grisebachstrasse 1a, 3400 Göttingen, Federal Republic of Germany.

Dr E. M. RIX, Royal Horticultural Society Gardens, Wisley, Ripley, Woking, Surrey, United Kingdom.

Dr G. ROSSMAN, Botanischer Garten der Ruhr-Universität, Hustadtring 81/402, 463 Bochum-Querenburg, Federal Republic of Germany.

Mr A. RØSVIK, Ringve Botanisk Hage, Universitet i Trondheim, 7000 Trondheim, Norway.

Professor H. B. RYCROFT, Kirstenbosch Botanic Garden, National Botanic Gardens of South Africa, Private Bag X7, Claremont 7735, South Africa.

Mr A. D. SCHILLING, Royal Botanic Gardens Kew, Wakehurst Place, Ardingly, Sussex, United Kingdom.

Mr P. N. D. SEYMOUR, Devonian Botanic Garden, University of Alberta, Edmonton, Alberta T6G 2E9, Canada.

Mr J. B. SIMMONS, Royal Botanic Gardens, Kew, Richmond, Surrey, United Kingdom.

Mr D. SMIT, Hortus Botanicus Vrije Universiteit, Van der Boechorststraat 8, Postbus 7161, Amsterdam, Netherlands.

Dr G. SMITH, Botanic Garden, Universiti Malaya, Lembah Pantai, Kuala Lumpur, Malaysia.

Dr M. C. SMITH, University Botanic Gardens, Department of Botany, Woodland Road, Bristol, United Kingdom.

Dr S. SNOGERUP, BCT Department, The Wallenberg Laboratory, Fack, 22007, Lund, Sweden.

*Dr E. SOEPADMO, Jabatan Botani, Universiti Malaya, Lembah Pantai, Kuala Lumpur 22-11, Malaysia.

*Dr L. ŠOMŠÁK, Botanická záhrada Prírodovedeckej fakulty Univerzity Komenského, Nábrežie armádneho generála L.Svobodu č. 11, 816-00 Bratislava, Czechoslovakia.

Dr ELÍAS R. de la SOTA, Facultad de Ciencias Naturales y Museo, Division Plantas Vasculares, Universidad Nacional de la Plata, Paseo del Bosque, 1900 La Plata, Argentina.

Professor W. T. STEARN, 17 High Park Road, Kew, Richmond, Surrey, United Kingdom.

Mr C. H. STIRTON, Pretoria National Botanic Garden, Botanical Research Institute, Private Bag X101, Pretoria, South Africa (presently South African Liaison Officer, Herbarium, Royal Botanic Gardens, Kew).

Dr A. L. STORK, Conservatoire et Jardins Botaniques, Case postale 60, 1292 Chambesy, Geneva, Switzerland.

Dr V. STRGAR, Biotechnical Faculty and Institute of Biology, University of Ljubljana, Izanska 15, 61000 Ljubljana, Yugoslavia.

Mr A. H. M. SYNGE, IUCN Threatened Plants Committee, c/o Royal Botanic Gardens, Kew, Richmond, Surrey, United Kingdom.

Mr N. P. TAYLOR, Royal Botanic Gardens, Kew, Richmond, Surrey, United Kingdom.

Mr P. THODAY, University of Bath, Calverton Down, Bath, United Kingdom.

Mr A. D. THOMPSON, 19 Pere and David Streets, Kitty, Georgetown, Guyana.

Dr P. A. THOMPSON, Royal Botanic Gardens Kew, Wakehurst Place, Ardingly, Sussex, United Kingdom.

Mr D. W. H. TOWNSEND, Royal Botanic Gardens, Kew, Richmond, Surrey, United Kingdom.

Monsieur P. VALCK, Jardins Botaniques de la Ville et de l'Université de Nancy, 36 rue Sainte Catherine, 54000 Nancy, France.

Mr J. A. VARLEY, Tropical Soils Unit, Land Resources Development Centre, Coley Park, Reading, United Kingdom.

Mr R. VIANE, Schouwvagerstrasse 16, 8000 Bruges, Belgium.

Mr A. P. VOVIDES, Instituto de Investigaciones Sobre Recursos Bióticos (INREB), Heroico Colegio Militar No 7, Xalapa, Veracruz, Mexico.

Mr S. WAHLBERG, World Wildlife Fund Sweden, Fituna, 140 41 Sorunda, Sweden.

Dr S. M. WALTERS, University Botanic Garden, Cambridge CB2 1JF, United Kingdom.

Mr J. F. WARRINGTON, Royal Botanic Gardens, Kew, Richmond, Surrey, United Kingdom.

Dr A. WEINSTEIN, Agricultural Research Organisation, Forestry Division 'Ilanot', Doar Na Lev Hasharon, Israel.

Mr D. WELLS, Nature Conservancy Council, PO Box 6, Godwin House, George Street, Huntingdon, United Kingdom.

Dr T. C. WHITMORE, Commonwealth Forestry Institute, South Parks Road, Oxford, United Kingdom.

Mr D. O. WIJNANDS, Botanische Tuinen van de Landbouwhogeschool, Generaal Foulkesweg 37, Wageningen, Netherlands.

Herr G. WINKEL, Schulverwaltung, Schulbiologiezentrum, Brockenweg 5A, 3000 Hannover 21, Federal Republic of Germany.

Dr B. A. WINTERHOLLER, Central Botanical Garden of the Kazakh Academy of Sciences, 480070 Alma-Ata 70, Kazakh SSR, USSR.

Mr J. A. WITT, University of Washington Arboretum, Seattle, Washington 98195, USA.

Mr J. W. WRIGLEY, National Botanic Gardens, PO Box 158, Canberra, ACT 2601, Australia.

* Contributors, but not present at the meeting.

The Council of Europe Resolution on Rare and Threatened Plants in Europe

The Committee of Ministers,

Referring to Resolutions no.2 of the European Ministerial Conferences on the Environment (Vienna 1973, Brussels 1976) on the protection of wildlife;

Having regard to the list of rare, threatened and endemic plants in Europe* commissioned by the European Committee for the Conservation of Nature and Natural Resources;

Recalling that man and all animals are dependent for their survival on the plant kingdom;

Recognizing that plants (species, subspecies, varieties, etc.) form a genetic resource of immeasurable value to mankind and that the economic potential of the plant kingdom is as yet only partly realized;

Recognizing the scientific, educational, recreational, aesthetic, cultural and ethical value of plants to mankind;

Noting that the list includes some 1,400 species as rare and/or threatened in Europe, of which more than 100 are in imminent danger of extinction and that the figure of 1,400 represents approximately one tenth of the total European flora;

Realizing that once a species becomes extinct, it cannot be recreated by man, and hence that it is of the utmost importance to ensure the conservation of as many species as possible for the economic, scientific and cultural benefit of mankind;

Recommends that the governments of member States of the Council of Europe be guided in their policy in this matter by the principles set out below;

1. Ensure adequate legal protection for all plants identified as endangered in the above-mentioned list with provision for licences to be issued for approved collection purposes;

2. Provide minimum legal protection for all plants against depredations not yet covered by law;

3. Institute or complete national surveys of plants that are rare or threatened within their boundaries for appropriate dissemination and publication. Such surveys should:

    (a) Include plants that are rare or threatened only in particular countries and therefore not included in the list;

* IUCN Threatened Plants Committee (1977). *List of rare, threatened and endemic plants in Europe*. Nature & Environment Series No. 14. Council of Europe, Strasbourg.

    (b) Identify the principal threats to the plants so listed;

    (c) Specify the action needed to ensure their survival;

4. Establish nature reserves and designate areas in which vegetation and flora are protected by law and stimulate the setting up of nature reserves by private bodies, with the long-term aim of ensuring that all species on the list can be found in such areas and in so doing contribute to the establishment of the European network of biogenetic reserves which was the subject of Resolution (76) 17;

5. Incorporate safeguards in future planning strategies to protect all species on the list, as the major threat to many plants is created by changing patterns of land use;

6. Stimulate, undertake and co-ordinate through competent organizations multidisciplinary research at national or international level, with particular emphasis on bringing together information on plants found in more than one country with a view to:

    (a) Extending and improving knowledge about the flora of those areas in Europe that are still insufficiently known botanically, so as to be able to make constructive proposals for conservation and planning purposes;

    (b) Promoting studies on the habitat, autoecology and population biology of each plant on the list to provide the information needed from which integrated conservation management plans can be formulated;

    (c) Promoting studies on the dynamics and ecology of the vegetation types in which the plants on the list occur;

7. Give appropriate support to scientifically based botanical gardens so that they have the facilities they need to propagate and grow the plants on the list and to distribute propagating material to other institutions and where appropriate re-introduce plants to the wild, with the aim of reducing the pressure on wild plant populations and at the same time drawing attention to the aesthetic, cultural and scientific importance of these plants;

8. Ratify for their States, if they have not already done so, the Convention on international trade in endangered species of wild fauna and flora, opened for signature in Washington on 3 March 1973;

9. Acknowledge that the plant kingdom is a dynamic system and needs to be monitored at stated intervals so that the list can be revised regularly;

10. Prepare and disseminate codes of conduct on rare and threatened plants;

11. Disseminate general information on the need to protect plants and on the protective measures set out in the European Committee's list.

The following note from *The IUCN Plant Red Data Book* (Lucas & Synge, 1978) may be relevant here:

"It is important to note that the adoption of all these measures in other parts of the world is not necessarily a realistic goal, although various parts of the recommendations could certainly be widely applied and incorporated in appropriate legislation. In many areas, the decline and extinction of plants are part of a much larger process of degradation in which the whole environment is affected by dramatic land use changes; here it is the total reassessment of development strategies that needs to be undertaken, involving the integration of conservation measures within development plans to assure the long-term survival of the region concerned. For instance, in most areas of tropical rain forest the scattered distribution patterns of species, in conjunction with threats endangering whole ecosystems, dictate different priorities: first and foremost, protection of as many diverse areas of forest as possible, with emphasis on larger areas which are easier to protect and more likely to be self-supporting without loss of diversity. Due to the imbalance of resources, particularly with regard to scientific expertise, it is highly desirable that wherever practicable such expertise should be made available to help in the development of strategies likely to ameliorate the situation. It is with this fundamental objective in mind that the IUCN Threatened Plants Committee has undertaken the task of data-gathering on a world scale, in the hope that it will serve to highlight areas where urgent action is most needed. Conservation must be built into long-term planning in these areas for, in both the short and the long term, it is upon the natural renewable resources such as forests and rangelands that their inhabitants will depend and must rely for continuing employment and income."

Agreed Resolutions of the 1975 Conservation Conference

on

The Function of Living Plant Collections in Conservation and in Conservation-Orientated Research and Public Education

Royal Botanic Gardens, Kew, 2–6 September 1975

1. This conference, conscious that the rich tropical floras of the world are now in great hazard, (1) urges that a strong network of nature reserves and conservation-orientated gardens should be established throughout the tropics both through the strengthening and development of existing foundations and through the creation of new ones where the need exists; (2) recommends that institutions in temperate countries should offer all possible help in this programme through technical aid, training and the secondment of personnel; and (3) urges that this aim should be pursued through the International Union for Conservation of Nature and Natural Resources to ensure good co-ordination and proper understanding of the importance of the work for the tropical countries themselves and for the whole of mankind.

2. This Conference urges that special attention be given to the Conservation of Threatened Floras particularly of Islands and those parts of the world with Mediterranean or similar climates since both are often inhabited by very large numbers of narrowly endemic species of plants endangered by human activities.

3. This Conference recommends that institutions maintaining plant collections (including seed collections) for conservation purposes should, in general, give priority to their local flora, so as (1) to benefit from local taxonomic, ecological, physiological and other pertinent specialist knowledge; (2) to reduce the need to simulate remote climates with the attendant costs and dependence on man-generated energy; (3) to be able to offer from direct experience information and advice concerning field conservation in the country of the institution, and (4) to provide a basis from which public interest and pride in the indigenous flora can be developed through display and education services.

4. This Conference urges all Governments to ratify the 'Convention on International Trade in Endangered Species of Wild Fauna and Flora' as soon as possible.

5. This Conference recommends that, wherever possible, all living plant collections grown for conservation purposes should also be stored in the form of seeds under appropriate conditions for long-term conservation.

6. This Conference urges that the propagation of rare and endangered species, including research into appropriate techniques, should be actively pursued by Botanic Gardens and other bodies maintaining living plant collections, and that such activities should be financially supported where necessary by Conservation, or other appropriate Organizations. Special attention should be given to economic plants and their wild relatives and to plants which are or might be commercially used.

7. This Conference urges that whenever threatened plants are taken into cultivation, this be done by means of seed and/or cuttings whenever possible so as not to deplete the wild populations.

8. This Conference, aware of the urgent need for scientifically verified lists of threatened species on a world scale, calls for the full support for the work of the IUCN Threatened Plants Committee in compiling such lists, and urges the task of propagating stocks of species on institutions maintaining living plant collections.

9. This Conference calls for the widest publicity to its full deliberations to be given in all appropriate quarters, and urges that the resolutions should be made available separately for this purpose with the minimum delay.

10. This Conference, being acutely aware of the urgency and complexity of many problems which have been raised during the sessions, urges the desirability of continued study and exchange of information, and the setting up of working parties to continue the study of outstanding issues, e.g.

    1. Listing of collections, documentation and dissemination of information
    2. Commercial use of wild species
    3. Preparation of codes of practise
    4. Publicity
    5. Relationship between institutions maintaining living plant collections and organizations concerned with nature conservation
    6. Compilation of a short list of rare and endangered plants of high scientific importance to be commended to botanic gardens to bring them into cultivation.

APPENDIX 3

The European List

Reprinted below is a short article by S. M. Walters that the editors feel provides a most useful summary of the conclusions that can be drawn from the *List of rare, threatened and endemic plants in Europe*, and so will be relevant to botanic gardens in their conservation rôles. Dr Walters is Chairman of the European Sub-Committee of the IUCN Threatened Plants Committee and one of the instigators of the European List, a subject about which he is uniquely qualified to write.

The article was originally published in *Naturopa* No. 31 (1978), the environmental magazine of the Council of Europe. The editors are most grateful to the Council for permission for this article to be reproduced here.

The publication of the 286-page *List of rare, threatened and endemic plants in Europe* (Council of Europe Nature and Environment Series 14, 1977) represents an important achievement, and one of which the many European botanists who voluntarily collaborated in its formulation may feel justly proud. The urgency and gravity of the problems facing plant conservation in Europe are, however, such that we cannot afford to be complacent. The 'List' is not an end in itself, but merely a tool to be used by and with governments and private organizations and individuals to plan practical action.

Its compilation is a co-operative project of experts from the *Flora Europaea* team and conservation organizations in different European countries by the Council of Europe, especially through its *ad hoc* working party set up in 1974, and by the newly-formed office of the Threatened Plants Committee of the International Union for Conservation of Nature and Natural Resources, housed at the Royal Botanic Gardens, Kew. From the latest analysis of the completed *Flora Europaea*, including the as yet unpublished final volume, we can say that there are in Europe about 11,500 different species of flowering plants and ferns, of which about 3,500 are endemic, i.e. occur nowhere outside Europe. The List contains about 2,100 species which are rare or threatened on a European scale, two thirds (1,400) of which are known to be rare or threatened on a world scale.

We should therefore be specially concerned about the fate of approximately one in every five vascular plants in Europe. Moreover, about 100 species are given the IUCN category 'Endangered'—plants which are in danger of total extinction and whose survival is unlikely if the causal factors continue to operate—and these are a matter of urgent concern for the conservation organizations in those European countries in which they are still to be found.

The listing of narrowly endemic species, i.e. those occurring in only one (European) country, enables us to see immediately a most significant fact—namely, that the floras of northern and central Europe are relatively poor in endemic species, whilst those of the Mediterranean countries are very rich. Greece (including Crete and the Aegean) is not much more than half the size of Great Britain, yet there are 676 Greek endemics compared with 15 in the British list! It is clear that, whatever co-operative effort we devise to protect the threatened flora of Europe, by far the greatest part of that effort should go into safeguarding the habitats of rare and local species in the countries of southern Europe. Much more international support is needed for the creation and effective maintenance of nature reserves and protected areas in the Mediterranean countries, and with it must go more scientific research into the rare Mediterranean plants and the vegetation in which they grow. In this connection the initiative of the recently-formed Organization for the Phyto-taxonomic Investigation of the Mediterranean Area (OPTIMA) in setting up a Conservation Commission is greatly to be welcomed.

One other generalization is important. The List enables us to say something at least about the kinds of species which are or were widespread in Europe (and may even extend outside Europe) but which seem to be declining simultaneously in many countries. There are in fact two very obvious groups: species of arable and disturbed ground (often called 'weeds' though many have no agricultural importance), and species of wetland and freshwater habitats. In both these cases, changes in agricultural methods and land use policies are responsible for the decline, and both groups need special attention and concern.

I want to conclude by listing what we, as conservationists concerned to protect the flora and vegetation of Europe, must see as urgent tasks. Now that we have the basic floristic information we must:

1. urge the organizations responsible for nature conservation in our own countries to give priority to the protection of those rare and threatened species listed as still occurring within their country by the creation of nature reserves, the enactment of appropriate legislation, the education of public opinion, and in any other way;

2. support the efforts of international bodies to continue the effective co-operation demonstrated in the production of this List, and in particular the scientific co-operation necessary to refine and enlarge our knowledge of the ecology of rare species in their natural habitats, so that we know how to protect them;

3. press for rapid progress towards a system of effective nature reserves for Europe which contain within them as many as possible of the listed rare and threatened species. It is essential to remember that we can ultimately only protect species by protecting the habitats in which they occur.

Index

Entries are omitted where they are evident from the title of the paper

Ornamental plants, 66, 69, 150, 151, 153, 156, 216
Orobanchaceae, 64
Orobus latyroides, 151
Orothamnus zeyheri, 130
Osmunda regalis, 114, 220
Osprey, 50
Ostrowskia magnifica, 149
Otanthus maritimus, 92
Oxalis acetosella, 137

Pacific, 20
Paeonia, 153
P. hybrida, 150
P. lagodechiana, 156
P. macrophylla, 156
P. majko, 156
P. mlokosewitschii, 156
P. wittmanniana, 156
Pal, B. P., 115
Palestine Sunbird, 48
Palmae, 64, 66, 67, 69, 70, 78, 202
Pamir mountain xerophytes, 143
Pancratium maritimum, 92, 93
Pandanaceae, 64
Pandion haliaetus, 50
Pansies, 167
Papaver, 78
P. bracteatum, 156
P. tenellum, 151
Paphiopedilum fairieanum, 114
P. insigne, 114
Parana Pine, 96
Parashorea, 65
Parodi, L.R., The Garden, 97
Parrots, 102
Pasque flower, 27
Pastinacopsis glacialis, 149
Patagonian-Andean forests, 96
Pauri Botanic Garden, 113
Pavetta, 66
Peatbogs, 58, 220
Pedicularis comosa, 109
Pehuéa, 96
Penang Water Fall Garden, 69
Pentaspadon, 66
Perring, F. H., 32, 35, 37
Persea lingue, 97

Petrea, 67
Petrosaviaceae, 64
Peucedanum arenarium, 110
P. palustre, 220
Phebalium, 103
Phenakospermum, 78
Philippines, 64, 68, 70
Phleum arenarium, 220
Phragmites, 138
Phyllagathis, 66
Physandra halimocnemis, 149
Phytophthora cinnamomi, 104
Pieniny mountains, 53, 54, 59
Pieniny National Park, 107
Pilularia globulifera, 21, 220
Pine, 136, 137, 138, 149, 215, 216
Pinguicula grandiflora, 220
Pinus, 96
P. cembra, 54
P. halepensis, 92
Plantago coronopus, 219
Platanus orientalis, 83
Platycerium, 113, 114
P. coronarium, 193
Pleione praecox, 114
Pneumatopteris callosa, 193
Podocarpus andinus, 97
P. nubigenus, 97
Polessje, 143
Polish Botanic Gardens, 60
Polish National Parks, 59
Polish rare plant species, 56, 57
Polish tree species, 57
Polish protected species, 61, 62
Polyalthia, 66
Polystichum braunii, 151
Polystichum lonchitis, 138, 151
Populus, 41, 96, 149, 215
P. nigra, 81
Port Blair Botanic Garden, 114
Potatoes, 200, 216
Potentilla silesiaca, 53
Prenanthes purpurea, 58
Primula, 80, 153
P. juliae, 156
P. veris, 27, 138
P. vulgaris, 138
P. woronowii, 156